NIFTY MATH NOTIONS

An Exploration of Puzzles, Problems,
Ideas, and Discoveries

NIFTY MATH NOTIONS

An Exploration of Puzzles, Problems, Ideas, and Discoveries

Charles L. Silver

SPIE PRESS
Bellingham, Washington USA

Library of Congress Cataloging-in-Publication Data

Names: Silver, Charles L., author.
Title: Nifty math notions : an exploration of puzzles, problems, ideas, and
  discoveries / Charles L. Silver.
Description: Bellingham, Washington : SPIE, [2024]
Identifiers: LCCN 2024007602 | ISBN 9781510674844 (paperback) | ISBN
  9781510674851 (pdf)
Subjects: LCSH: Mathematical recreations. | Mathematics–Anecdotes.
Classification: LCC QA95 .S494 2024 | DDC 793.74–dc23/eng/20240314
LC record available at https://lccn.loc.gov/2024007602

Published by
SPIE
P.O. Box 10
Bellingham, Washington 98227-0010 USA
Phone: +1 360.676.3290
Fax: +1 360.647.1445
Email: books@spie.org
Web: www.spie.org

The content of this book reflects the work and thought of the author. Every effort has
been made to publish reliable and accurate information herein, but the publisher is
not responsible for the validity of the information or for any outcomes resulting from
reliance thereon.

Cover image credit: Ernesto Mora

Printed in the United States of America.
First printing 2024.
For updates to this book, visit http://spie.org and type "PM375" in the search field.

Contents

Acknowledgments

I first would like to thank John Lester Miller for his recommendation to write this book. Next, I'd like to thank Errol Morris of Fourth Floor Productions for allowing his staff to help me with the book. My thanks also go to Fabiola Washburn for her helpful revisions. Most of all, I wish to thank Josh Kearney for months of improvements and rewrites. I could not have completed the book without his expert help. Finally, I would like to thank my wife Susan, who persisted in encouraging me to complete the manuscript as well as suggesting improvements.

Preface

While strolling in my neighborhood one day, I saw a woman digging in her garden. She looked up at me, said something, and we began a conversation. One thing she said was that her daughter, who was standing right there, was good at mathematics. I looked down at the tiny girl. She couldn't have been more than five years old. I asked her whether she liked math and she nodded. I asked her what she was working on right now. She said her multiplication tables. I asked her what was $7 \cdot 9$ and she looked befuddled. Then she said, "I'm not good yet at my 9-times tables." She, her mother, and I were all embarrassed, so I changed the topic. Soon, though, her mother returned to math. She said her daughter really loved playing with numbers, but her kindergarten teacher discouraged it, saying that it just wasn't normal for a girl of her age.

Later, while continuing my stroll, I wondered how I could have helped the girl. One thing that occurred to me was to ask her whether she knew what $6 \cdot 9$ was. If she did, then I could have told her that $7 \cdot 9$ is just one more 9 than $6 \cdot 9$ (i.e., $54 + 9$). I was pretty sure she was good at adding, and I thought she might be able to do that. What also seemed helpful to say was that 9 times any number always gives digits that themselves add up to 9. For example, $2 \cdot 9 = 18$, and the sum of 1 and 8 is 9.

I could also have said, "What's $2 \cdot 9$?" I figured she surely knew that, and would have brightened up and said, "That's easy, $2 \cdot 9 = 18$." Then I could have said, "Notice that the digits add up to 9; that is, $1 + 8 = 9$." I could then tell her this always works with 9s. To "prove" it, I could have asked, "What's $3 \cdot 9$?" Again, I think she would have felt proud knowing that the answer is 27. I could then have repeated that the two digits $2 + 7$ add up to 9 once again, and I could have told her again that this always works.

Then, maybe I could have asked her to add 9 to 27. With a little thought, I believe she would have said 36. I could then say that 36 is $4 \cdot 9$, 9 more than $3 \cdot 9$, just like before. Then I could have told her to notice that the first digit, 3 of 36, is one more than the first digit, 2 of 27. Hopefully, I could have followed that by telling her to notice that the second digit is always one less. So, 2 • 9 is 18 and 3 • 9 is 27. She might not have gotten it immediately. So, I could have said: "Look, the 9-times tables are easy if you first add 1 to the first digit and take away 1 from the second as you go from step to step."

She might still be somewhat puzzled. So, I could have repeated, "Start with 18, then to find out what 3 • 9 is, add 1 to 1, which is 2, and then take away 1 from 8, which gives you 7. Putting the two digits together gives you 27, which is 3 • 9." Then I could have repeated myself again, saying that to get 4 • 9, just add 1 to the 2 of 27, and take away 1 from the 7, and you get 36.

Surely, I might have had to say more to her about 9s and maybe other numeric relations, and at some point, I could have told her to write these numbers down and look at them: 18 (2 • 9), 27 (3 • 9), 36 (4 • 9), 45 (5 • 9), 54 (6 • 9), 63 (7 • 9). Then (hoping I'd been at least partially successful), I could have repeated: "You see, at first you didn't know how to multiply 7 • 9, and now you do, if you know what 6 • 9 is. If you don't, you can use your fingers to count from 18 to 27 to 36 to 45 to 54 and finally to 63. And there are many more interesting aspects to 9."

The way the number 9 works is "nifty" and could have helped the girl learn arithmetic faster. More complex nifty notions occurred to me, many that might appeal to older audiences, and I decided to put them in a book. Some of them are very simple, and others are harder and sometimes more technical. I enjoy those that I chose, and I hope readers find them appealing as well.

This work is a pastiche—a blend of theory, history, musings, anecdotes, and challenges. Many of the notions and problems I explore are well beyond the scope of primary or secondary school curricula. All the same, I tried to make them comprehensible to audiences of all ages and backgrounds, and to provide plenty of curious diversions along the way.

To those ends, you'll encounter more than 40 numbered exercises, all of which have at least partial solutions in the back of the book. There are some additional suggested exercises too, but these aren't numbered and don't have solutions. As the book progresses, the material becomes more challenging, often building on the ideas explored in previous chapters. But it is my hope that any reader can enjoy the book, whether dipping in at random or moving through methodically.

The most prolific mathematician of the 20th century, Paul Erdős, used to say that the most elegant proofs were in "God's Book" or simply in "The Book" (since Erdős was an atheist). It is hoped that some of the proofs and exercises in this book are elegant enough to make it into "The Book."

Charles L. Silver
July 2024

Chapter 1
Paul Erdős

When Paul Erdős was only three years old, he could calculate the number of seconds a person had lived. At 19, he elegantly simplified a complex proof by the Russian mathematician Pafnuty Chebyshev. In honor of his result, mathematicians recited this rhyme[1] on Erdős' behalf:

Chebyshev said it, and I say it again,
There's always a prime between n and 2n.

Figure 1.1 Photograph of Paul Erdős (1913–1996). Photo credit: George Csicsery.

Paul Erdős was jobless and homeless most of his life. He wandered the earth as a vagabond for half a century, inspiring mathematicians in more than 25 countries to solve interesting problems, both old and new. He would arrive at a mathematician's house and announce, "my brain is open."[2] Living mostly on coffee, he worked on mathematics 19 hours a day, seven days a week. He claimed, "a mathematician is a machine for turning coffee into theorems."[1]

In Fig. 1.2, Erdős discusses math with a precocious child named Terence Tao. Now an adult, Tao is widely considered to be the most brilliant mathematician alive.

Figure 1.2   Photograph of Paul Erdős and 10-year-old Terence Tao. Photo credit: Terence Tao.

Possibly what's best known about Paul Erdős is the Erdős number. That number reflects the degree of connectedness of a mathematician to a publication by Erdős. Erdős himself has Erdős number (EN) 0, having published papers with himself. If a mathematician has authored a paper with Erdős (but is not Erdős), then his EN is 1. If a mathematician has authored a paper with someone with EN 1 (but does not have an EN of 0 or 1), then his EN is 2. The boy in the above photo, Terence Tao, has an EN of 2. He authored a paper with Gergely Harcos and one with Vitaly Bergelson, both of whom coauthored with Erdős.

The existence of the Erdős number was one of the influences in the creation of the Kevin Bacon number, or just the Bacon number, which measures the number of degrees of separation an actor or actress is from Bacon, in terms of having acted in the same film. Kevin Bacon seems to have been selected for this honor by virtue of so many other actors having been in films with him, and many having been in films with those in films with him, and so on and so forth.

However, numerous other actors have acted in more films than Bacon and have a tighter degree of connectedness than he does. On some websites it is claimed that more than 454 actors are better connected. In mathematics, by contrast, Paul Erdős published more papers than anyone else in history: more than 1525 of them. Thus, the honorary Erdős number seems to be an even better measure of connectivity than its Hollywood equivalent.

The Dinner Party Problem

Ramsey's theorem comes out of combinatorial mathematics and graph theory. But we can explain it simply with a famous example about arithmetic series.

Imagine the simplest arithmetic series you can: 1, 2, 3, 4, . . ., N. What about 2, 4, 6, 8, . . ., N? This is also an arithmetic series, but in this one each number differs from its predecessor by 2. What Ramsey's theorem tells us is that if we split a set like $\{1, . . ., N\}$ into two parts, one of them will contain an arithmetic series of, say, length 7. We could state one of Ramsey's results as follows: "For every k, there exists a number N such that if we split the set $\{1, . . ., N\}$ into two parts, one of the parts contains an arithmetic series of length k."

This is weird and counterintuitive.

Paul Erdős delved into generalizations of problems like this. His explanation makes the results from Ramsey's theorem a bit more comprehensible.

Suppose that six people go to a dinner party. Ramsey's theorem states that at least three people all know each other or three are mutual strangers.

To show that six people at a party is sufficient for three persons knowing each other or three persons not knowing each other, we will draw a hexagon, i.e., a six-sided figure (Fig. 1.3). The six vertices represent the six persons at the party. We will color red those lines representing people who mutually know each other and blue for those who mutually don't.

If, no matter how you connect the vertices, you wind up with an all red-colored triangle or a blue-colored triangle, then three people must know each other or three persons do not. This will prove that six people are sufficient.

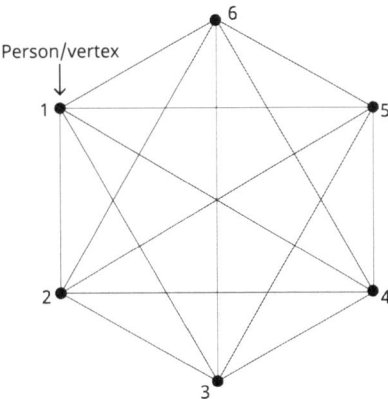

Figure 1.3 Illustration of the dinner party problem with all lines in black. Image credit: Ernesto Mora.

Take Person (vertex) 1 and connect the lines to the five other Persons (vertices). At least three of the five lines must be red or blue. Suppose, then, that Person 1 knows Persons 2, 4, and 5 (Fig. 1.4).

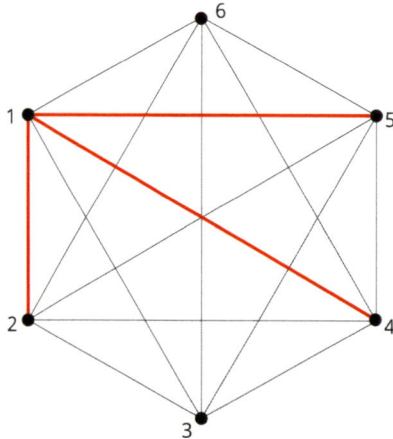

Figure 1.4 Illustration of the dinner party problem with three lines in red. Image credit: Ernesto Mora.

Color lines 1–2, 1–4, and 1–5 red, as shown above. Consider Person 2 and Person 5. If they know each other, the line 2–5 would be red, and a red triangle would be formed. So, we must color 2–5 blue (Fig. 1.5).

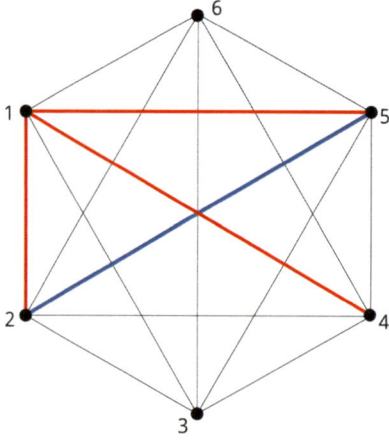

Figure 1.5 Illustration of the dinner party problem with three red lines and one blue line. Image credit: Ernesto Mora.

The same situation holds between Person 4 and Person 5. If they're connected by red, then a red triangle is formed, showing that Persons 1–4–5 know each other. So, to avoid that, we color 4–5 blue (Fig. 1.6).

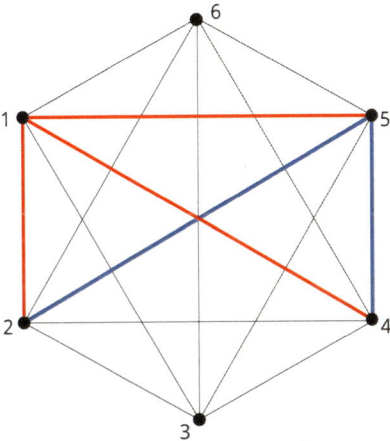

Figure 1.6   Illustration of the dinner party problem with three red lines and two blue lines. Image credit: Ernesto Mora.

Now, we're in a quandary: if we color 2–4 red, we end up with a red triangle, 1–2–4. But if we color 2–4 blue, we end up with a blue triangle: 2–4–5. We must draw a colored line between 2 and 4. But now, no matter how we color it, we form a red triangle or a blue one, proving an instance of Ramsey's theorem, which is what we set out to prove.

With six people (vertices), there are 15 lines. We can see this just by counting them. But there's a nifty formula that will tell us how many connecting lines there are for any number of vertices. That formula is $n \cdot (n-1)/2$. Taking our six-sided figure, $6 \cdot 5/2 = 15$. But there are 32,768 (2^{15}) possible ways of coloring these lines red or blue.

Extending the party problem, Erdős considers how many people must be at a party so that there are four people who either must know each other, or four who are mutual strangers. For four people, there are 153 lines and 2^{153} ways to color these lines red or blue. The answer is 18, but Erdős tells us that it's not so simple as before. How many people do you need to arrive at five such people? Nobody knows. The number is between 43 and 55. Erdős imagines that there's an evil spirit, he calls him the "Supreme Fascist," who tells us he will exterminate the human race if we can't determine the number of people for five such people. Erdős counsels us to work diligently on the problem, using both mathematics and computers. For six people, we should try to destroy the Supreme Fascist before he destroys us; we couldn't solve it for six people.

Chapter 2
Young Gauss

Figure 2.1 Potrait of Carl Friedrich Gauss (1777–1855). Image credit: Österreichische Nationalbibliothek, Wien, Germany.

Carl Friedrich Gauss made so many contributions to so many mathematical fields that he's referred to as "the prince of mathematicians." Gauss may even have been the greatest mathematician of all time. Surely, he was one of the most precocious.

When he was three years old, Gauss corrected the figure his father had arrived at in calculating a payroll. He is supposed to have said, "Father, the reckoning is wrong, it should be..."[3] Later in life, Gauss would joke that he learned to calculate before he could read.

When Gauss was only seven (or eight, or ten, depending on whom you read), he attended a class taught by J. G. Büttner, who was somewhat of an ogre. Apparently wanting to keep his students busy so they wouldn't disturb him, Büttner asked them to add all the numbers from 1 to 100—that is, to add $1 + 2 + 3 + \ldots + 100$. That should take them quite some time, Büttner thought. He was therefore quite surprised when young Gauss immediately rose from his seat and handed in his slate containing only the single number 5050. The rest of the students toiled for an hour before arriving at their answers, adding their slates to the pile. Later in life, Gauss boasted that his answer was the only correct one.

According to many accounts, Gauss' method was the first to put the series of numbers to be added on one line, starting with 1 and ending at 100. Then, underneath this series he wrote the series written backwards, like this:

$$
\begin{array}{ccccccccccccc}
1 & + & 2 & + & 3 & + & \ldots & + & 98 & + & 99 & + & 100 \\
100 & + & 99 & + & 98 & + & \ldots & + & 3 & + & 2 & + & 1 \\
\hline
101 & + & 101 & + & 101 & + & \ldots & + & 101 & + & 101 & + & 101
\end{array}
$$

Looking at the numbers to be added, we can see that the sum of the two series is 100 times $101 = 10{,}100$. Since the above pair of series represents two sums from 1 to 100 and we want only one sum and not two, we take 50 (1/2 of 100) times 101 to arrive at 5050. 5050 is the single number Gauss wrote on his slate, placing it on his teacher's desk almost as soon as Büttner had posed the problem, saying in his peasant German dialect, "*Ligget se'*": "There it lies."[3]

It turns out that Büttner was not a complete ogre. When he saw how gifted young Gauss was, he bought him an advanced arithmetic text to study on his own. He also introduced Gauss to a mathematics tutor, a young man of 20 named Johann Martin Bartels, who had been teaching himself mathematics. Soon, Bartels realized that Gauss needed something more than he could provide, so he presented him to Professor Eberhard August Wilhelm von Zimmerman at a nearby college.

I've read the above description of Gauss' method numerous times. But I have never seen any evidence that this was the way Gauss solved the problem. I have doubts, since others cite a simpler method that is much more transparent. Just add the first number to the last, the second to the next-to-last, the third number to the third-to-last, and so on—i.e., $1 + 100$, $2 + 99$, $3 + 98$, $\ldots$. See below:

$$\begin{array}{ccccccccccc}
1 & + & 2 & + & 3 & + & \ldots & + & 48 & + & 49 & + & 50 \\
100 & + & 99 & + & 98 & + & \ldots & + & 53 & + & 52 & + & 51 \\
\hline
101 & + & 101 & + & 101 & + & \ldots & + & 101 & + & 101 & + & 101
\end{array}$$

50 times $101 = 5050$ immediately, without needing to add larger numbers and then divide by two. Gauss may have even intuited this more general formula: $S_n = n/2(a_1 + a_n)$, where S_n is the sum of any n consecutive numbers, a_1 is the first number, and a_n is the nth.

The more complicated method, which can be considered a form of mathematical induction (the topic of our next chapter), might also have been the one used by young Gauss. As it turns out, this method yields a nice generalization, which can be helpful for any mathematical series.

A mathematical series is one in which a fixed number is added to each succeeding number. In Gauss' problem, the first number is 1, and 1 is added to each subsequent number. However, in a mathematical series, in general, 1 isn't necessarily the first number, and though the subsequent numbers each differ by the same quantity, that quantity doesn't need to be 1. For example, in the series 2, 5, 8, 11, 14..., 2 is the first number, and we arrive at each subsequent number by the addition of 3. The difference, which we'll call "d," between each number in this series is 3. Now that we have d, we can generate a neat formula, which Gauss may have also immediately intuited.

Consider this summation:

$$\begin{array}{rl}
S_n = & a + \quad\quad 0d + a + \quad\quad 1d + \ldots + a + (n-1)d \\
S_n = & a + (n-1)d + a + (n-2)d + \ldots + a + \quad\quad 0d \\
\hline
2S_n = & 2a + (n-1)d + 2a + (n-1)d + \ldots + 2a + (n-1)d
\end{array}$$

Take 1/2 the sum (1/2 times $2S_n$), which is $n[2a + (n-1)d]/2$. This nicely generalizes the first (somewhat more complicated) method above for summing the first hundred numbers. In the classroom, Gauss solved the problem when the beginning number a was 1; the difference between each number d was 1; and there were 100 numbers in the series. But to sum any mathematical series S_n when a, d, and n are known, use $S_n = n[2a + (n-1)d]/2$.

Gauss' formula, $S_n = n[2a + (n-1)d]/2$, has four variables, S_n, a, n, and d. Given any three dependent variables, one can find the value of the fourth—the independent—variable. (However, note that finding n can be in some cases a bit trickier, since to solve for n you may run into n^2, and then need to solve a quadratic equation.)

Exercise 1: Let a be 3 instead of 1. And let d be 7 added to each subsequent number until we reach eight numbers. What's their sum, S_8?

Exercise 2: Just for fun, suppose that Gauss was asked a slightly different question: "How many numbers in a series would you need to reach 565 (S_n), where the first number a is 7, then continuing to add some large d, say, 53?" We can use the identical formula, except in this case it's n that's the unknown (or independent) variable. And, as mentioned, finding n can be tricky. But it's pretty nifty that Gauss' idea can lead to four different sorts of problems, depending on which variable you select as the independent one: n, S_n, a, or d.

It doesn't end there. Gauss' formula can be used to solve problems that are sometimes solved using mathematical induction. For instance, adding successive odd numbers always yields a perfect square: $1 + 3 + 5 + \ldots + 2n - 1 = n^2$. For any series like this, little Gauss' formula hands us the answer almost immediately.

Look at a simple example using the formula $S_n = n[2a + (n - 1)d]/2$, where $n = 50$, $a = 1$, and $d = 2$. We should get n^2, which in this case is 2500:

$$S_{50} = 50[2 + (50 - 1)(2)]/2 = 50[2 + (49)(2)]/2 = 50(2 + 98)/2$$
$$= 50(100)/2 = 2500.$$

Chapter 3
Induction and Recursion

Before we go any further, we need to explain mathematical induction. Mathematical induction is a hypothesis used to prove that something true in one case is true in infinitely many cases. The idea is often illustrated by a depiction of falling dominoes (Fig. 3.1).

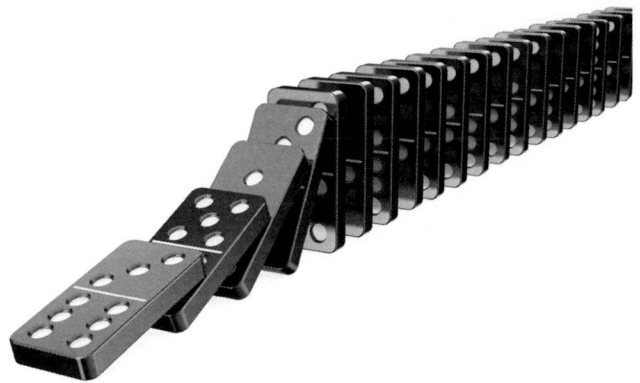

Figure 3.1 Photograph of dominoes falling. Photo credit: AlexLMX / Shutterstock.com.

According to mathematical induction, if an initial domino has a given property P, and if we assume that when any domino k has a property P, it passes that property on to domino $k + 1$—i.e., the next domino in line—then all dominoes have P.

The simplest form of mathematical induction infers that a statement involving some natural number n holds for all values of n. The proof consists of two steps:

1. Base Case: Prove that the statement holds for 0, or for 1.
2. Induction Hypothesis (IH): Prove that for every n, if the statement holds for n, then it holds for $n + 1$. In other words, assume that the

statement holds for some natural number of your choosing n, and prove that it holds for $n + 1$.

Now we can show Gauss' result more generally, using mathematical induction.

PROOF:
To Prove: For any natural number $n \geq 1$: $1 + 3 + 5 + \ldots + 2n - 1 = n^2$.
Base Case: Let $n = 1$. $2(1) - 1 = 1 = 1^2$.
IH: For any natural number k: $1 + 3 + 5 + \ldots + 2k - 1 = k^2$.
To Prove: $1 + 3 + 5 + \ldots + 2k - 1 + 2(k + 1) - 1 = (k + 1)^2$.
Proof: $k^2 + 2(k + 1) - 1 = (k + 1)^2$ (using induction hypothesis).
$\quad k^2 + 2k + 2 - 1 = k^2 + 2k + 1 = (k + 1)^2$.

An anonymous poet provides us with another glimpse at proof by induction:

A woman in liquor production
Owns a still of exquisite production.
The alcohol boils
Through magnetic coils.
She says that it's "proof by induction."

But who first introduced the method of proof by mathematical induction? Some say Euclid used mathematical induction in his proof that there are infinitely many primes (or, more precisely, that the number of primes is unending). Formally, however, the method of induction was first used by Francesco Maurolico (Fig. 3.2) in his book, *Arithmeticorum libri duo* (1575). Maurolico demonstrated that the sum of the first n odd numbers of any series equals the nth square number. Or, as formulated and proved above, that for $n \geq 1$: $1 + 3 + 5 + \ldots + 2n - 1 = n^2$. How do we know that Maurolico used induction in his proof? We know this because he first proved the "base case" and a few other initial cases when $n \geq 1$. Then for the inductive step, he used the proposition that the nth square number plus the following odd number equals the following square number.

Blaise Pascal in his *Traité du triangle arithmétique* (1665) explicitly formulated mathematical induction—i.e., he formulated it as a principle of proof. This was an advance from Maurolico, who used induction in his proof but did not formulate it as an actual proof principle. However, in one of Pascal's letters he acknowledged Maurolico's prior use of the method.

Figure 3.2 Portrait of Francesco Maurolico (1494–1575). Line engraving by M. Bovis after Polidoro da Caravaggio / Wellcome Images licensed under CC BY 4.0 International.

Statement of mathematical induction:

    a. Prove that an initial number, called the base case, has property P.

    b. Assume, by the induction hypothesis, that property P holds for some arbitrary number k and then:

    c. Prove that $k + 1$ has the property as well.

    d. Therefore, every number greater than or equal to the base case has property P.

Let's try an example. Prove that for any integer $n \geq 1$, $n^3 + 2n$ is evenly divisible by 3.

First prove the base case, that $1^3 + 2(1)$ is evenly divisible by 3.

Then, (b) assume that for arbitrary positive $k \geq 1$, $k^3 + 2(k)$ is evenly divisible by 3. This assumption is called the induction hypothesis, abbreviated as IH.

After assuming the IH, we prove that the assertion is also true for the formula $(k + 1)^3 + 2(k + 1)$, where $k + 1$ is substituted for k in $k^3 + 2(k)$.

After proving that $(k + 1)^3 + 2(k + 1)$ is also evenly divisible by 3, the proof is finished. We proved by induction that for any positive integer $n \geq 1$, $n^3 + 2n$ is also divisible by 3, which is what we wanted to prove.

> PROOF:
> To Prove: For any integer $n \geq 1$, $n^3 + 2n$ is evenly divisible by 3.
> Base Case: $n = 1$. $1^3 + 2(1) = 3$, and 3 is divisible by 3.
> IH: Assume that $k^3 + 2k$ is evenly divisible by 3.
> To Prove: $(k + 1)^3 + 2(k + 1)$ is also divisible by 3.
> Proof: $(k + 1)^3 + 2(k + 1) = (k + 1)(k + 1)^2 + 2k + 2 = (k + 1)(k^2 + 2k + 1)$
> $+ 2k + 2$.
> $= (k^3 + 2k^2 + k) + (k^2 + 2k + 1) + 2k + 2$.
> $= (k^3 + 2k) + (3k^2 + 3k + 3)$.

Since we know by assumption—i.e., the induction hypothesis—that $k^3 + 2k$ leaves no remainder when divided by 3, we know there's some integer q such that $3q = (k^3 + 2k)$. Furthermore, $3k^2 + 3k + 3 = 3(k^2 + k + 3)$. So, $(k^3 + 2k) + (3k^2 + 3k + 3) = 3q + 3(k^2 + k + 1)$, which $= 3(q + k^2 + k + 1)$, which when divided by $3 = q + k^2 + k + 1$, and we're finished.

Thus, we've shown by mathematical induction that any number of the form $n^3 + 2n$ is evenly divisible by 3.

What is nifty about mathematical induction is that it permits you to prove infinitely many mathematical facts with just a few finite steps—actually just two (three if you count the induction hypothesis).

Exercise 3: The standard, initial example of mathematical induction usually presented in textbooks is the proof that $1 + 2 + 3 + \ldots + n = [n(n + 1)]/2$. Using the method demonstrated above, try to prove this on your own. Remember, solutions for this and the following exercises can be found in the back of the book.

Exercise 4: Prove for $n \geq 4$: $3^n > 2n^2 + 3n$.

Exercise 5: Here's an interesting induction problem. This one's not so straight-forward as the one above. In fact, it's rather tricky.

> PROOF:
> To Prove: That 21 divides $4^{n+1} + 5^{2n-1}$ for $n \geq 1$.
> Base Case: For $n = 1$, check $4^{1+1} + 5^{2(1)-1} = 4^2 + 5^1 = 16 + 5 = 21$.
> IH: Assume that $4^{k+1} + 5^{2k-1}$ is divisible by 21. Now, try to prove it
> for $k + 1$.

To Prove: $4^{(k+1)+1} + 5^{2(k+1)-1}$ is divisible by 21. Simplifying, try to prove that $4^{k+2} + 5^{2k+1}$ is divisible by 21 (knowing by IH that $4^{k+1} + 5^{2k-1}$ is divisible by 21).
Proof: ?????

Before continuing, I wish to mention a theorem used often to prove other things. It's a simple theorem, so simple that proving it by mathematical induction seems like overkill. But, if you want a mathematically acceptable proof, you need induction.

First, I'll explain the theorem with a simple metaphor. Pretend you've uprooted a tree and turned it upside down. The ball of roots represents the tree's top node. The tree has infinitely many nodes from finitely many branches. Stop and think a moment. How could there be infinitely many nodes and only finitely many branches? Clearly (I think), there must be an infinite path wending its way through nodes on the finitely many branches.

To speak a little more technically, we're interested in Kőnig's lemma (a lemma is a mathematical result that is typically used to prove something else, usually a theorem). The root is the top node of a tree, and all finitely many nodes directly connected to the top node are its children, and they may in turn have children as well. The single node from which children spring is called their parent.

The lemma states that if T is a tree with infinitely many nodes, but every node has only finitely many children, then there must be a path that continues from parent to child with infinitely many nodes along its path (i.e., with infinitely many descendants).

To prove this formally requires induction. What we do is take the first node (the root), which by hypothesis has infinitely many descendants. Then we inductively select a child of that root that itself has infinitely many descendants. (We know we can do this, since there are infinitely many nodes.) We continue this process of inductively choosing a child from each parent such that the child has infinitely many descendants, and we're done.

Exercise 6: We all know (don't we?) that a set of n elements has 2^n subsets. There are many ways to show this. As an exercise, prove this fact by mathematical induction.

Here's an unusual and surprising theorem that's nifty indeed. Since this theorem is a surprising one, its proof requires more than the usual fiddling around. But I guess such a surprising theorem *should* require more work than usual. Try your hand at proving Exercise 7.

Exercise 7: Using mathematical induction, prove that $1^3 + 2^3 + 3^3 + \ldots + n^3 = n^2(n + 1)^2/4$.

As an illustration of how proofs do and don't work, there's a problem that is featured in many math texts: prove for $n \geq 4$ that $n! > 2^n$. Let's look at this one together before you try to finish it on your own.

Exercise 8: Prove for $n \geq 4$ that $n! > 2^n$.

There's a simple issue here. Take $n = 3$ (which is less than the first number of the problem). $n! = 3! = 1 \cdot 2 \cdot 3 = 6$, whereas $2^3 = 8$. Here we took 3 as an example, and in this example $3! < 2^3$. But now take $n = 4$. Then $4! = 24$, whereas 2^4 is only 16. The difference between them is 8. For $n = 5$, $5! = 120$, and 2^5 is only 32. The difference between them has grown to 88! By continuing in this way, we can see that the difference is growing so fast that proving $n! > 2^n$ might seem like a waste of time. This doesn't eliminate the need for a real proof.

More directly, after $n = 4$, subsequent multipliers for $n!$ (5, 6, 7, ...) become greater and greater than just multiplying by 2s. That is, the multipliers for $n!$ where $n \geq 4$ keep increasing, while those for 2^n remain at 2. Isn't this proof enough? Well, no. But why is this? Why is a mathematical proof so important when the result is completely obvious without proof? This concern relates to our fondness for the Greek invention of proofs, before which the Egyptians actually knew many mathematical relationships, such as instances of the Pythagorean theorem, but did not develop the notion of proof.

So, now that we require proofs even when the result is obvious, see if you can provide proofs of your own for Exercise 8, as well as for the following exercises.

Exercise 9: Prove that $7^n - 1$ is divisible by 6 for each positive integer n.

Exercise 10: Prove that $7^n + 5$ is divisible by 6 for each positive integer n.

Exercise 11: Prove by induction for $x \geq 2$, $n \geq 1$, that $x^n - 1$ is divisible by $x - 1$.

Exercise 12: Prove $2^n > n^2$ for $n > 4$. If you get stuck, first prove (a): for $n > 2$, $2^n > 2n + 1$. Then, use (a) to help prove (b): $2^n > n^2$ for $n > 4$.

Recursive Definitions: From the Finite to the Infinite

Recursive definitions are explained here because of their neat property of allowing us to go from the finite to the infinite in simple, easy steps. Mathematical induction also goes from the finite to the infinite in only two small steps. I hope you'll agree that taking two small steps from the finite to the infinite is definitely nifty.

The simplest way (though perhaps not the most revealing) to explain what a recursive definition is—or for that matter a recursive anything—is "something that recurs," where "recurs" just means "repeats itself." Or maybe one should just say that recursive definitions continue to refer to each other. (An advisory point: there are several distinct kinds of recursion, not simply the kind we're covering. Entire books have been written on the subject.)

Possibly the most frequently used, simple recursive definition is the two-line recursive definition of the natural numbers, where S is short for "the successor of" (where the successor of n is the next number after n, $n + 1$):

1. 0 is a natural number.
2. If x is a natural number, so is $S(x)$.

In other words, once we know that 0 is a natural number, we can get the successor of 0—i.e., $S(0)$, which is 1. Once we have 1, we can take its successor and get 2, then 3, and so on. The "and so on" means up to the infinite (not actually stretching to an infinite number, but to every finite number). So, in just two steps we've accounted for every natural number. I think that's fairly astonishing and hope you do too.

Suppose you wish to add $SSSSS(x)$ to $SSSSSSSSS(x)$. What do you get? The obvious solution is to just add the two bunches of S's. There are five S's in the first sequence, and nine in the second. Five plus nine equals 14; so, there are 14 successors of x. But there's just a little bit more to say about this. We've arrived at the natural number 14 based on thinking from clause 1 that 0 is the first natural number. Suppose that clause $1'$ were to say that 20 is the first natural number. Then adding $SSSSS(x)$ to $SSSSSSSSS(x)$ would equal $5 + 20 + 9 + 20$, which equals 54. But then of course we wouldn't know that the first 19 numbers are natural numbers. If we invent clause 3 that says if x and y are natural numbers, then so is $x + y$; thus we've created addition. And if clause 4 says that we can add additions as many times as we want, we get multiplication, and then exponentiation... and any other operation on the natural numbers. It all starts with taking the successor of 0.

Chapter 4
Euclid

Figure 4.1 Portrait of Euclid (ca. 325–265 BCE). Woodcut from *Les Vrais Pourtraits et Vies des Hommes Illustrés Grecz, Latins, et Payens* by André Thevet (1584).

Euclid's Proof of the "Infinity" of Primes

Paul Erdős considered Euclid's proof of the infinity of primes to be in "The Book." He's not alone. It's long been thought one of the niftiest proofs in all of mathematics.

In many math texts it is claimed that Euclid's proof is an indirect one, a proof by contradiction, a *reductio ad absurdum*. In a *reductio* proof, it is first assumed that the negation of some statement ϕ (phi) is correct, and then later in the proof a contradiction is reached, proving that φ is really in fact correct. However, as can be seen below, Euclid's proof makes no such assumption and does not arrive at any contradiction. In other words, his proof is a direct one. It can be found in Euclid's *The Elements*, Book IX, Proposition 20.[4]

> PROPOSITION 20: Prime numbers are more than any assigned multitude of prime numbers.

It may be of some interest to look first at Sir Thomas Heath's formulation of Euclid's proof in his translation and commentary on Euclid's *The Elements*. It is quoted below.[4]

> We have here the important Proposition that the number of prime numbers is infinite.
> The proof will be seen to be the same as that given in our algebraic textbooks. Let a, b, c, ..., k be any prime numbers.
> Take the product $abc... k$ and add unity.
> Then $(abc... k + 1)$ is either a prime number or not a prime number.
>   1. If it is, we have added another prime number to those given.
>   2. If it is not, it must be measured by some prime number, say p.
> Now p cannot be identical to any of the prime numbers a, b, c, ..., k.
> Therefore, since it divides $(abc... k + 1)$ also, it will measure the difference of unity:
> which is impossible.
> Therefore, in any case, we have obtained one fresh prime number.
> And the process can be carried on to any extent.

A small quibble with Heath's summary of Euclid's proof is that Euclid never uses the word "infinite" in his proof, preferring to say that prime numbers can be found that exceed "any assigned multitude of prime numbers." Euclid's formulation would be accepted by anyone who doesn't believe in a completed infinity of numbers yet accepts that there is no last number. Such a person would demur accepting Heath's formulation of Euclid's proof, since that proof invokes the concept of an actual infinity.

So, "Who cares?" you may ask. Aristotle, for one. According to him, "Now if what we have been saying establishes the kind of infinity which depends on division, but makes it look as though there cannot be an infinite by addition—at least not one that exceeds every definite magnitude—that would be another reasonable outcome of the discussion."[5] Thus, according to Aristotle, a unit can be infinitely divided, but an infinite sum is not possible. It seems then that to Aristotle, Euclid's own proof is acceptable, but Heath's reformulation of it is not.

Even the prince of mathematicians, Carl Friedrich Gauss, expressed a "horror of the actual infinite." He wrote, "I protest against the use of an infinite magnitude as something completed.... Infinity is merely a way of speaking."[3] So, Gauss too would not have accepted Heath's version of Euclid's proof.

Here is a modern formulation of Euclid's proof (notice again that the proof is a direct one). Unabashedly, in contradistinction to Aristotle's and Gauss' views, this proof does invoke an actual infinity.

> To Prove: There are infinitely many primes.
> Given any sequence of primes starting with 2, show that there's a prime not in the sequence.
> Let $p_1, p_2, \ldots, p_n$ be such a sequence of primes.
> Take $(p_1 \cdot p_2 \cdot \ldots \cdot p_n) + 1$.
> For any prime p_q in this sequence, dividing p_q into $(p_1 \cdot p_2 \cdot \ldots \cdot p_n) + 1$ will leave a remainder of 1.
> Thus, $(p_1 \cdot p_2 \cdot \ldots \cdot p_n) + 1$ must itself be a new prime or a composite number with at least one new prime as a factor (i.e., a prime number larger than p_n).

Exercise 13: Check that each one of $(2 + 1)$, $(2 \cdot 3 + 1)$, and $(2 \cdot 3 \cdot 5 + 1)$ is itself prime. Does continued multiplication of these primes plus 1 always reach a new prime? If not, where exactly does the first non-prime—i.e., the first composite—show up in $(p_1 \cdot p_2 \cdot \ldots \cdot p_n) + 1$, starting with 2 as p_1?

Exercise 14: Give an alternative proof to Euclid's proof of the infinity of primes by assuming that the largest prime is p and using $N = p! + 1$.

The largest known prime is $2^{82,589,933} - 1$, found on December 7, 2018 by Patrick Laroche, a participant in the Great Internet Mersenne Prime Search (GIMPS). When written in base ten, this number has 24,862,048 digits. Any number that can be written as one less than a power of two is called a Mersenne prime. Even when n is prime, $2^n - 1$ is not always prime. For instance, $2^{11} - 1 = 23 \cdot 89$. The last eight largest known primes on record are Mersenne primes, which are easier to obtain than other sorts of prime numbers. No one knows whether there are larger non-Mersenne primes.

The Pythagorean Theorem

Euclid's most famous proof is his proof of the Pythagorean theorem. Before delving into Euclid's own proof, let's examine some of the numerous so-called "alternative proofs" of the theorem of Pythagoras since it's so nifty. There are possibly 370, perhaps only 367, of such proofs.

As we know, the point of the theorem is to prove that the sum of the squares on two sides of a right triangle equals the square of the hypotenuse.

In the construction in Fig. 4.2, let lowercase a and b stand for the two sides of a right triangle, and let lowercase c stand for the hypotenuse. Thus, we want to prove that $a^2 + b^2 = c^2$.

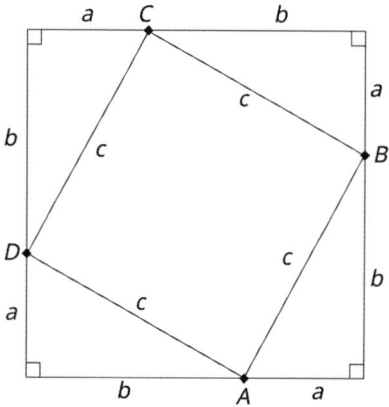

Figure 4.2 Square within a square (an alternative proof of Pythagoras' theorem). Image credit: Ernesto Mora.

The area of the outer square $= (a + b) \bullet (a + b) = a^2 + 2ab + b^2$.
The area of the inner square $= c^2$.
The area of a single right triangle whose two sides are a and $b = 1/2(ab)$.
Thus, the area of all four triangles $= 4[(1/2)(ab)] = 2ab$.
Therefore, the area of the inner square $c^2 =$ the area of the outer square, minus the area of the four triangles. So, $c^2 = (a^2 + 2ab + b^2) - 2ab$, which is to say $c^2 = a^2 + b^2$, and we're done.

Before looking at Euclid's proof, we will feature another spiffy alternative proof of the Pythagorean theorem, this one produced by James Garfield, the 20th president of the United States. Garfield arrived at his proof before becoming president, when he was a member of Congress in 1876.

Look at Fig. 4.3.

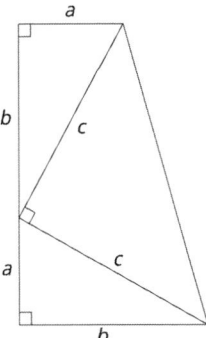

Figure 4.3 James Garfield's proof of Pythagoras' theorem. Image credit: Ernesto Mora.

Begin to construct the above figure by starting out with the bottom right triangle formed by the sides a and b, and the hypotenuse c. Then flip the triangle, reconstructing it so that its horizontal length is a and the vertical height is b. Its hypotenuse is likewise c. Then, by connecting the diagonal line from side a on top to side b on the bottom, we get a trapezoid. The area of the trapezoid equals 1/2 the sum of its bases times the height. In other words, $A = 1/2(a + b) \cdot (a + b) = (a + b)^2/2$. Another way to get the area is to add the areas of the separate figures. That is, $2 \cdot 1/2(ab) + 1/2c^2$. (The two sides c intersect at a right angle, since the other two adjacent angles are complementary.)

After these clever proofs, why should we bother with Euclid's own proof?

Some people dislike Euclid's proof, primarily because they find it difficult to understand and because the elements used to secure the proof seem so complicated and roundabout. It may well be the most difficult proof in Euclid's entire book, but it's possibly also the most interesting. Students throughout the ages have wrestled with trying to understand it, often giving up in frustration.

The renowned philosopher Arthur Schopenhauer, himself highly skilled in the art of reasoning, protested that this proof, "rather than instruct the student, could easily overwhelm him: 'Lines are drawn, we know not why, and it afterwards appears that they were traps which, close unexpectedly and take prisoner the assent of the astonished reader.'"[6]

Sir Thomas Heath has a much different take on Euclid's proof. Heath comments that the Pythagorean theorem in Euclid's Book I "is effected by a construction and proof so extraordinarily ingenious [that it] is a veritable *tour de force* which compels admiration, notwithstanding the ignorant strictures of

Schopenhauer who... calls Euclid's proof 'a mouse-trap proof' and 'a proof walking on stilts, nay a mean, underhanded proof.'"[7]

Not only is the proof "a veritable *tour de force*," but it's nifty as well. What's nifty is that the proof fits in a place where seemingly so little information is available. The proof is difficult precisely because it so ingeniously employs scanty information in surprising ways. The above, simpler proofs—one with a tilted square inside another square, and the other a trapezoid containing two equivalent triangles—though ingenious, are not genuine alternative proofs at all; neither one can be squeezed in after Proposition 46 in *The Elements*.

To make Euclid's proof seem less mysterious, it seems a good idea to first present a simpler and somewhat enigmatic proposition on which his proof is partially based. Let us consider Euclid's statement of Proposition 41.[7]

> PROPOSITION 41: If a parallelogram has a base that's the same for a triangle and is in the same parallels, the parallelogram is double the triangle.

Now, consult Fig. 4.4.

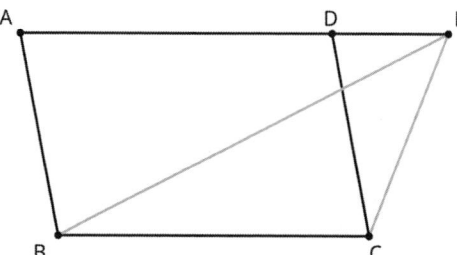

Figure 4.4 Parallelogram illustrating Euclid's Proposition 41. Image credit: Ernesto Mora.

Euclid writes: "I say that the parallelogram ABCD is double the triangle BEC."[7]

We know that the area of a parallelogram equals its base times its height. And we also know that the area of a triangle equals 1/2 its base times its height. Being "in the same parallels" means that the parallelogram ABCD and the triangle BEC have the same height. Now, since they also have the same base, the area of ABCD must be twice the area of triangle BEC. Voilà, the result is now clear.

With Proposition 41 out of the way, we're better able to follow Euclid's proof of the theorem of Pythagoras, as the justifications for other propositions are less obscure than for Proposition 41.

Euclid's Proof of the Pythagorean Theorem

To me, one of the most surprising aspects of Euclid's proof is what appears to be his central insight regarding a seemingly unimportant parallel line drawn from the vertex of the triangle's right angle to the base of the square on its hypotenuse. This parallel line nicely separates the hypotenuse's square into two rectangles, each of which equals the area of one of the two squares on the other two triangle's bases. This is what needs to be shown.

How remarkably insightful that parallel line turns out to be! Much later, once ratios and proportions have been introduced in Euclid's *The Elements*, its significance is obvious, but it's far from obvious at the point where Euclid produces his Pythagorean theorem, pulling it out of a hat so to speak.

PROPOSITION 47:
To Prove: In right triangles, the square of the hypotenuse is equal to the sum of the squares of the other two sides.
Let ABC be a right triangle with right angle BAC.
Prove that the square BC is equal to the sum of the squares on BA and AC.

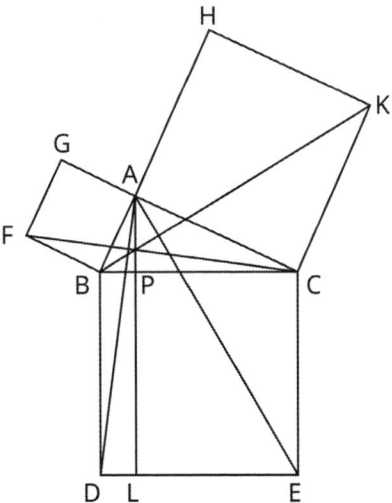

Figure 4.5 Diagram of Euclid's proof of Pythagoras' theorem. Image credit: Ernesto Mora.

First, the right triangle ABC needs to be constructed (Fig. 4.5), where the right angle is at A. Now, we're ready to prove that $BC^2 = AB^2 + AC^2$.

Construct on BC the square BDEC, and on BA and AC, construct the squares GB and HC, respectively. (Note that Euclid uses two diagonal points to label rectangles.)

Next, construct from point A a line AL parallel to BD, and join AD and FC. (Euclid does not label it, but let P be the point where AL intersects BC.)

Then, since each of the angles BAC and BAG is a right angle, it follows that with a straight line BA, and at the point A on it, the two straight lines AC and AG not lying on the same side make the adjacent angles equal to two right angles; therefore, CA is in a straight line with AG. (Euclid proved this already in Proposition 14.)

For the same reason, BA is in a straight line with AH.

And since angle DBC is equal to angle FBA, for each is a right angle, let the angle ABC be added to each.

Therefore, the whole angle DBA equals the whole angle FBC.

And since DB equals BC and FB equals BA, the two sides AB and BD equal the two sides FB and BC, respectively.

Thus, the triangle ABD is congruent to the triangle FBC.

Now, the parallelogram BL is double the area of triangle ABD, for they have the same base BD and are in the same parallels BD and AL. And the square BGAF is double the area of triangle FBC, for they have the same base FB and are in the same parallels FB and GC. (This is what was shown in Proposition 41, above.)

But doubles of equals are equal to each other. Therefore, the parallelogram BL is equal in area to the square HC.

The second half of the proof is identical to the first half, presented above. It consists in showing that the other parallelogram LECP is equal in area to the other square ACKH. But we won't bother the reader by repeating the steps above to do the exact same proof with different letters.

The Pythagorean theorem is hereby proved.

Figure 4.6 Portrait of Pythagoras (ca. 570–495 BCE). Illustration by J. Augustus Knapp from *The Secret Teachings of All Ages* by Manly P. Hall (1926).

What about Pythagoras' own proof? Is it completely different from Euclid's?

Though Pythagoras is best known for the theorem bearing his name, he is not thought to have proved it, nor to have proved any other theorem for that matter. Early Egyptians regularly used the Pythagorean triple 3, 4, and 5, but there's no indication that they'd proven anything either.

Pythagoras is said to have founded a religious sect consisting of mystics and number worshippers who attributed all their own numerological discoveries to Pythagoras. One famous discovery, that the $\sqrt{2}$ is irrational (i.e., not a fraction), is said to have been arrived at by Pythagoras' follower Hippasus of Metapontum. Legend has it that when Hippasus announced that the $\sqrt{2}$ is irrational while aboard a ship of Pythagoreans, he was immediately tossed overboard and drowned. Since according to the Pythagoreans' creed everything in the world can be reduced to rational numbers, it's easy to see why Hippasus' result would have been an annoyance.

The Pythagorean society adopted certain principles of life, such as secrecy, vegetarianism, refusal to eat beans, never urinating toward the sun, celibacy, belief in reincarnation and the transmigration of souls—the passing of a person's soul after death either to another human body, to that of an animal, or to an inanimate body. This cycle of rebirth is eternal unless somehow the soul is released. Legend has it that while walking one day, Pythagoras encountered a man beating a dog. Recognizing the voice of his deceased friend in the dog's whimpering, Pythagoras prevailed upon the man to stop beating him.

Other Pythagorean beliefs include the following views: that reality is mathematical, philosophy is used for spiritual purification, the soul is divine, and that various symbolic patterns possess mystical significance. One surprising element of Pythagoreanism is that both men and women were permitted to become members. Several women Pythagoreans became noted philosophers, which was remarkable for the time.

In the words of Proclus, the most authoritative philosopher of late antiquity:

> If we listen to those who wish to recount ancient history, we may find some of them referring this theorem to Pythagoras and saying that he sacrificed an ox in honour of his discovery. But for my part, while I admire those who first observed the truth of this theorem, I marvel more at the writer of *The Elements*, not only because he made it fast by a most lucid demonstration, but because he compelled assent to [a generalization of it].[8]

So, we're left with Euclid's marvelous proof of the theorem of Pythagoras in the axiomatic system of geometry spelled out in Euclid's *The Elements*. It is agreed that Euclid's system is an early forerunner of modern axiomatic systems. But how close to modern axiomatizations is it?

Chapter 5
Axioms

The Axiomatic Method

The axiomatic method is a procedure by which an entire system of knowledge or science (e.g., geometry) is generated in accordance with specified rules by logical deduction from certain basic propositions, which in turn are constructed from a few things taken to be primitive. In his book *The Elements*, Euclid axiomatized geometry in a way that approaches the axiomatic systems of today, but he was still far from satisfying modern standards.

The definition of "primitive" is a bit of a moving target. A modern axiomatic system begins with only a list of axioms, *not* with these three notions: (1) definitions, (2) postulates, and (3) common notions, all of which are found in Euclid. From these three primitive notions, Euclid deduces numerous propositions, using any deductive reasoning that seems appropriate to him. Nowadays, theorems are proved from axioms alone, using only a few primitive, formal inference rules. Customarily, just a single rule is used: *modus ponens*.

Modus ponens (MP) is the most important rule of inference for an axiomatic system. MP allows you to write the sentence Q on a line of a proof if the sentences P and $P \rightarrow Q$ (which can be read: "if P, then Q") occur on previous lines.

In *The Elements*, Euclid sometimes uses questionable reasoning patterns. For example, he uses the unusual (and suspicious) proof method called "superposition," which can be viewed uncharitably as one geometric figure flying in the air and then landing on top of another. Superposition can be seen in Proposition 4, where Euclid proves two triangles congruent by side-angle-side (SAS). Furthermore, Euclid's propositions do not always follow from his primitive notions using even his *own* methods. For instance, in the very first proposition, where Euclid shows that an equilateral triangle can be constructed on a straight line, he assumes that two intersecting circles can touch each other at two given points, which is not obtained by any type of reasoning. He seems just to assume this.

To illustrate what a modern axiomatic system looks like, we present an important, modern axiomatic system of numbers, which will come in handy in the chapters ahead. It would probably seem better at this juncture to present an axiomatic system specific to geometry, since we could then compare it directly to Euclid's. Perhaps the best of these modern axiomatizations of geometry is due to Alfred Tarski (Fig. 5.1), especially since Tarski proves that his system is "complete." Unfortunately, it is difficult to "read off" the geometric content of such modern systems as Tarski's. Consequently, we turn to axiomatizations of other mathematical theories whose axioms clearly reflect what they are about.

Figure 5.1   Photograph of Alfred Tarski (1901–1983). Photo credit: George M. Bergman from Archives of the Mathematisches Forschungsinstitut Oberwolfach.

Basic Logical Notation

Before we go any further into axioms, we need to familiarize ourselves with the language of logic. It's a simple language, as you can see, but you may want to flag these pages so that you can easily refresh your memory. Below is a basic set of symbols used in logic, alongside their corresponding ordinary language translations in italics. I'll introduce a few more in later chapters, but this group is the most important.

$\wedge$	*And*
$\vee$	*Or*
$\rightarrow$	*If. . ., then. . .*

↔	*If and only if...* (sometimes written "iff")
¬	*Not*
∴	*Therefore*
∀x	*For all x...* (also known as "the universal quantifier")
∃x	*For some x...*, or, *There exists an x such that...* (also known as the "existential quantifier")
φ (phi) and ψ (psi)	(Variables: their use is defined on a case-by-case basis.)

Peano Arithmetic (PA)

Here's a transparent axiomatic theory of the natural numbers (beginning with 0), called "Peano arithmetic," or PA for short. Peano's axioms (from which PA derives) are named for the late 19th century Italian mathematician Giuseppe Peano. He gets the credit, but most of Peano's work was just a recapitulation of work done by the German mathematician Richard Dedekind. For these Dedekind–Peano axioms, the only inference rule that we need is *modus ponens* (MP).

In the axiomatization of PA, S of something means the successor of that something. For instance, $S(x)$ means the successor—the next number—after x. More concretely, $S(0) = 1$, $S(1) = 2$, and so on. Also, $SS(0) = 2$.

PA begins with familiar axioms of equality, often included in the underlying logic itself and not mentioned separately.

Equality Axioms:

E1. $x = x$	Identity (Id)
E2. $x = y \rightarrow y = x$	Symmetry (Symm)
E3. $(x = y \land y = z) \rightarrow x = z$	Transitivity (Trans)

Peano Axioms:
1. $0 \neq S(x)$ (Kleene's system reverses this to $S(x) \neq 0$.)
2. $S(x) = S(y) \rightarrow x = y$
3. $x + 0 = x$
4. $x + S(y) = S(x + y)$
5. $x \cdot 0 = 0$
6. $x \cdot S(y) = (x \cdot y) + x$
7. $[\varphi(0) \land \forall x(\varphi(x) \rightarrow \varphi(Sx))] \rightarrow \forall x\varphi$

The φ in Peano axiom 7 is a metavariable, making it an axiom schema ranging over infinitely many predicates, which means that PA is an infinite axiomatization. The other six axioms range only over numbers.

One way to convert PA to a finite axiomatization is to substitute, in place of axiom 7, the axiom $x \neq 0 \rightarrow \exists y[x = S(y)]$. The system with 7' replacing 7 is called Q due to Raphael Robinson (the husband of Julia Robinson, whom we will read about later). With this replacement, many theorems that can be proved in PA cannot be proved in Q. However, Q is still very important, for it is strong enough to be shown incomplete by Gödel's famous incompleteness result.

Here are two very basic theorems of PA that *cannot* be proved in Q.

> Theorem of PA, *not* provable in Q: $0 + x = x$.
> Theorem of PA, *not* provable in Q: $S(x) \neq x$.

However, both $0 + x = x$ and $S(x) \neq x$ *are* provable in PA, essentially using Peano axioms 3, 4, and 7 (assuming the equality axioms).

The main point of exhibiting PA and Q is to contrast such systems with that of Euclid. Euclid's system of geometry, as described above, has several varieties of primitive notions and informal inference rules, whereas a modern system has only formal axioms and inference rules.

Chapter 6
Angles of a Triangle

Here's a geometry problem known as "the hardest easy geometry problem." It's easy because it can be solved by ordinary means, constructions, and inferences. It's hard because in order to solve it, many constructions are needed. It's also called "Langley's Adventitious Angles" after Edward Mann Langley, who posed this problem in *The Mathematical Gazette* in 1922.

Given isosceles triangle ABC in Fig. 6.1, with ∠B and ∠C both equal to 80° and ∠A = 20°...

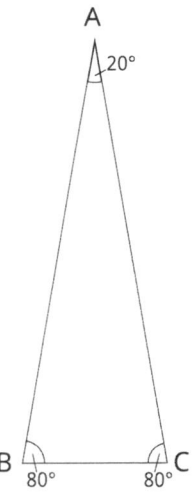

Figure 6.1 Diagram of Langley's adventitious angles: black. Image credit: Ernesto Mora.

Draw the extra lines in the following order: first draw BE, making ∠CBE = 60° and making ∠EBA = 20°.

Then, draw CF, making ∠BCF = 50° and ∠ECF = 30°.

Connect E to F in Fig. 6.2.

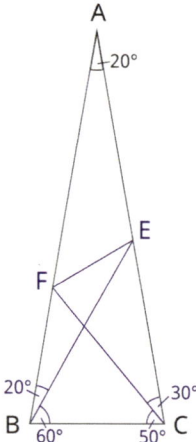

Figure 6.2 Diagram of Langley's adventitious angles: black, blue. Image credit: Ernesto Mora.

The problem is to find ∠BEF.

Construction:

Draw BG such that ∠CBG = 20°.

Since 20° + 80° = 100°, ∠CGB = 80° and BC = BG (triangle CGB is isosceles).

Therefore, ∠BGE = 100° (since a straight line = 180°).

In triangle BGE: ∠BEG = 40° (sum of the angles of a triangle = 180°).

Therefore, BG = GE (triangle BGE is isosceles).

Now, look at triangle CBF: ∠BFC = 50°. Therefore, BC = BF (triangle CBF is isosceles).

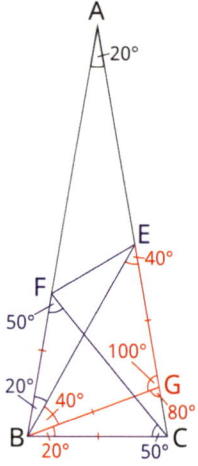

Figure 6.3 Diagram of Langley's adventitious angles: black, blue, red. Image credit: Ernesto Mora.

Connect G and F.

Look at triangle BFG. We already know that BG = BF and that the central angle = 60°. ∠BGF = ∠BFG = (180° − 60°)/2 = 60° (since they must sum to 180°).

Therefore, triangle BFG is equilateral, and GF = BF.

Since GF = GE, ∠GFE = ∠GEF = (180° − 40°)/2 = 70°.

Therefore, since 40° + ∠GEF = 70°, ∠BEF must be 30°.

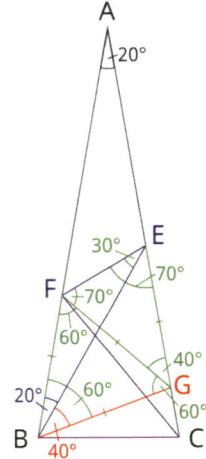

Figure 6.4 Diagram of Langley's adventitious angle: black, blue, red, green. Image credit: Ernesto Mora.

An easy geometry problem is to prove that the angles of a triangle sum up to 180°.

Figure 6.5 shows triangle ABC.

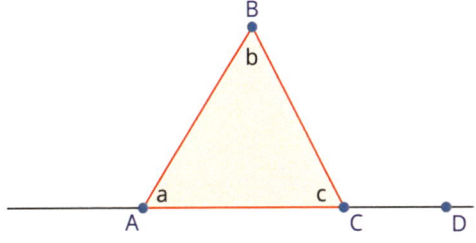

Figure 6.5 Triangle ABC. Image credit: Ernesto Mora.

Figure 6.6 is one proof (without words).

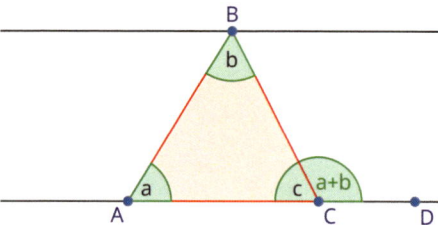

Figure 6.6 First proof that triangle ABC's angles = 180°. Image credit: Ernesto Mora.

Figure 6.7 shows a second way.

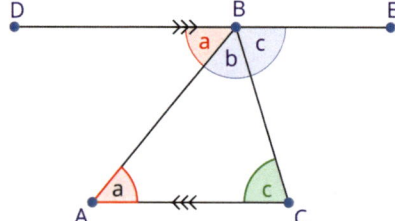

Figure 6.7 Second proof that triangle ABC's angles = 180°. Image credit: Ernesto Mora.

Try to prove it a third way on your own.

Chapter 7
Aristotle

In ancient times, even before Euclid, Aristotle formulated early patterns of systematic inference (though the Stoics, before him, had some rules of their own).

Figure 7.1 Photograph of a bust of Aristotle (384–322 BCE). Bust by Lysippos (ca. 330 BCE) / Ludovisi Collection / Jastrow / public domain.

Aristotle is of course well known for being a philosopher, and somewhat less well known as Plato's pupil and as teacher of Alexander the Great. Among Aristotle's many accomplishments is his square of opposition, which is briefly covered below. Aristotelian logic was so influential that it was adopted throughout the medieval period in history. Even as late as the year 1800,

Immanuel Kant considered Aristotelian logic complete and believed that no new logical discoveries could improve it.

Figure 7.2 shows Aristotle's square of opposition.

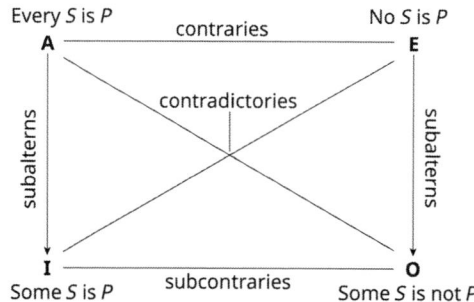

Figure 7.2   Diagram of Aristotle's square of opposition. Image credit: Ernesto Mora.

As can be seen in the figure, "Every *S* is *P*" is an example of an A-proposition. We saw this sort of statement denoted by ∀*x* in Chapter 5. It can also be rendered as "All *S* are *P*." Or in more modern terms, "for all *x*, if *x* is *S*, then *x* is *P*."

One of Aristotle's valid syllogisms is AAA, whereby it is meant that if every *S* is *P*, and every *P* is *Q*, then every *S* is *Q* (deliberately glossing over distinctions). Here is one of Aristotle's examples of a particular AAA ("B<u>A</u>rb<u>A</u>r<u>A</u>") syllogism:

> All men are mortal.
> All Greeks are men.
> Therefore, all Greeks are mortal.

What can we say about this syllogism? By common sense alone, it seems right. But what do we actually mean when we say, "it seems right"?

The two initial statements are called the "premises." The third statement is called the "conclusion." Considering them altogether, there are two questions we must ask ourselves. *If* the premises are true, must the conclusion be true as well—i.e., does the conclusion follow from the premises? For the syllogism above, the answer is yes. The second question is: *Are* the premises true? Again, the answer for the syllogism above is yes.

The first question is the equivalent of asking whether the argument is "valid." An argument is considered valid in the event that *if* all the premises are true, then the conclusion must be true as well.

The second question asks whether the argument is "sound." An argument is sound when not only is it valid, but the premises *are* actually true.

The syllogism above is both valid and sound. The conclusion does follow if the premises are true, and the premises are true. But you should know that an argument can be valid without being sound. It cannot, however, be sound without being valid.

Now that we're up to speed on validity and soundness, let's look at the most famous of Aristotle's syllogisms:

> All men are mortal.
> Socrates is a man.
> Therefore, Socrates is mortal.

There's one problem here. Aristotle made no room in his syllogisms for singular propositions. "Socrates is a man" is such a proposition. Thus, surprisingly, this famous example is not a valid syllogism in Aristotle's system. It has been suggested that the second premise could be considered an A-proposition if it were changed to "All Socrates are men," but that too was shunned by Aristotle, though accepted in "Aristotelian logic," the logic of the Middle Ages, which began as the logic of Aristotle but was supplemented by other principles.

For instance, the 13th century logician Peter of Spain employed singular propositions in his work the *Tractatus*. Thus, he'd have accepted "Socrates is a man" in a syllogism. Tracing the development of all principles in Aristotle's logic that were supplemented by others throughout the Middle Ages would be a Herculean task. There are books on the subject, but to my knowledge, none is comprehensive.

We've looked at Aristotle's famous AAA syllogism; there are many others. For instance, here's an AOO (refer back to the square of opposition if you've forgotten what "O" represents):

> All philosophers are thinkers.
> Some thinkers cannot be understood.
> Therefore, some philosophers cannot be understood.

Let's look at this in modern symbolic form. (Don't let the new notation confuse you. *Ua* symbolizes "can be understood"; $\neg Ua$ symbolizes "cannot be understood.")

$\forall x(Px \rightarrow Tx)$ All philosophers are thinkers.
$\exists x(Tx \wedge \neg Ux)$ Some thinkers cannot be understood.
$\therefore \exists x(Px \wedge \neg Ux)$ Therefore, some philosophers cannot be understood.

Can we prove the conclusion above from its premises? Actually, a proof within a formal system is called a "derivation." So, can we derive that conclusion from those premises?

Here's a purported derivation—with inferences unexplained.

1. $\forall x(Px \rightarrow Tx)$
2. $\exists x(Tx \wedge \neg Ux)$
To Prove: $\exists x(Px \wedge \neg Ux)$
3. $Ta \wedge \neg Ua$ from 2
4. $Pa \rightarrow Ta$ from 1
5. $Pa \wedge \neg Ua$ from 3 and 4
6. $\exists x(Px \wedge \neg Ux)$ from 5

Did you think that this derivation was valid? It's faulty: the syllogism is *invalid*. Here's one way to spot the logical error: though all philosophers are thinkers (by assumption, anyway), there may also be thinkers who are not philosophers. If we suppose that only these non-philosophical thinkers cannot be understood, then we don't arrive validly at the conclusion that there are some philosophers who cannot be understood. Another counterexample can be found by assuming $\forall x(Px \rightarrow Tx)$ to be "vacuously true"—i.e., that all philosophers are thinkers, but that there are no philosophers.

We can set up an alternative syllogism similar to the one above, except that this one *is* valid:

$\forall x(Tx \rightarrow Px)$ All thinkers are philosophers.
$\exists x(Tx \wedge \neg Ux)$ Some thinkers cannot be understood.
$\therefore \exists x(Px \wedge \neg Ux)$ Therefore, some philosophers cannot be understood.

Here's a (correct) derivation of the conclusion from the premises:

1. $\forall x(Tx \rightarrow Px)$
2. $\exists x(Tx \wedge \neg Ux)$
To Prove: $\exists x(Px \wedge \neg Ux)$
3. $Ta \wedge \neg Ua$ from 2
4. $Ta \rightarrow Pa$ from 1
5. $Pa \wedge \neg Ua$ from 3 and 4
6. $\exists x(Px \wedge \neg Ux)$ from 5

What's the difference between the two derivations? Why is this latter one valid? Are all premises true in both? No, the first premise in the latter one is false: not all thinkers are philosophers, which means that the syllogism is not sound. But that does not make the conclusion false. It only tells us that it's not guaranteed to be true by the validity of the argument.

To better understand vacuous truth, let's return to the square of opposition. As in the derivations above, the A-proposition can be symbolized in modern logical terms as $\forall x(Sx \rightarrow Px)$, "every S is P": for all x, Sx is Px. From this, its subaltern—$\exists x(Sx \wedge Px)$, "some S is P": for some x, Sx is Px—can be deduced in Aristotelian logic. Let's look closer at this deduction.

Make S be "___ is in the empty set." Make P be "___ is greater than 2." Then $\forall x(Sx \rightarrow Px)$ means "All elements in the empty set are greater than 2," but it does not follow from this that some elements in the empty set ($\varnothing$) are true, since there are no elements in $\varnothing$. So, anything we say about what's in $\varnothing$ is true —that it's not greater than 2, or even that it is greater than 2. It is said to be true vacuously.

For a more concrete example, take the sentence "Every son of John is well behaved." Now, suppose that John has no sons. It's vacuously true that all of them are well behaved, and it's also vacuously true that they're not well behaved—they all *misbehave*. Anything said of John's children is true vacuously.

If a person in real life were to say that all of John's children are well behaved, the listener would normally infer that John has children. This points to the fact that some mathematical inferences are not always identical to inferences made in non-mathematical contexts.

To render in logical terms that all of John's children are well behaved and that John has children, we would symbolize it this way: $\forall x(Sx \rightarrow Px) \wedge \exists x(Sx)$, where Sx stands for "child of John" and Px stands for "well behaved." Thus, if there are some S's, Aristotle's A-proposition does imply his I-proposition (look at the subaltern arrow from the A-proposition to the I-proposition).

In modern terms, Aristotle's propositions carry existential import, meaning that for him there must be S's—the class of S's must be non-empty. Most modern mathematicians eschew Aristotle's square of opposition, since to them $\forall x(Sx \rightarrow Px)$ does *not* carry existential import.

Another limitation to Aristotle's square is that it considers only monadic predicates, where modern logicians deal with polyadic predicates such as *Fxy*. (A monadic predicate takes a single variable *x*.)

There are other matters of Aristotle's logical system that are considered faulty or incomplete by standards of modern logic. But there's no contesting his genius in laying a foundation for logic that sufficed for many centuries, only a tiny fragment of which has been mentioned here.

Chapter 8
Riddles

Figure 8.1 Photograph of Charles Lutwidge Dodgson (1832–1898), better known by pen name Lewis Carroll. Photo credit: UK National Media Museum / John Cummings / Flickr Commons / public domain.

Lewis Carroll, famous for his Alice books (*Alice's Adventures in Wonderland* and *Through the Looking-Glass*), was an accomplished mathematician, a logician, and a church deacon. Among many, many other achievements, Lewis Carroll is known for his humorous syllogisms. Here are a few of Carroll's puzzles,[9] designed to help the logic student arrive at valid conclusions. See whether you can solve them:

1. All babies are illogical.
2. Nobody is despised who can manage a crocodile.
3. Illogical persons are despised.
∴ ?????

One valid conclusion is: no one who can manage a crocodile is a baby.

There are many ways to analyze the above argument. Here's a way that appeals to me and I hope to you. Translating the conclusion slightly ("logicizing" it): we want to show that for any person x, if x can manage a croc, then x is *not* a baby. So, take an arbitrary person p, who can manage a croc. The task then is to show that p is not a baby. Well, p can't be despised, according to premise 2. And, if p is not despised, then p is not illogical, by premise 3. And, by premise 1, p must then *not* be a baby. Voilà. Babies cannot manage crocodiles. Surprising, right?

Exercise 15: Here is a more complex puzzle of Carroll's. A solution can be found in the back of the book.

1. None of the unnoticed things met with at sea are mermaids.
2. Things entered in the log as met with at sea are sure to be worth remembering.
3. I have never met with anything worth remembering when on a voyage.
4. Things met with at sea that are noticed are sure to be recorded in the log.
∴ ?????

Exercise 16: If the last exercise wasn't too hard, try this one:

1. The only animals in this house are cats.
2. Every animal is suitable for a pet that loves to gaze at the moon.
3. When I detest an animal, I avoid it.
4. No animals are carnivorous unless they prowl at night.
5. No cat fails to kill mice.
6. No animals ever take to me except those in this house.
7. Kangaroos are not suitable for pets.
8. None but carnivora kill mice.
9. I detest animals that do not take to me.
10. Animals that prowl at night always love to gaze at the moon.
∴ ?????

As we know, Carroll wrote several humorous poems. Here's one of his poetic riddles, followed by its equally poetic solution:[10]

John gave his brother James a box:
About it there were many locks.
James woke and said it gave him pain;

So gave it back to John again.
The box was not with lid supplied,
Yet caused two lids to open wide:
And all these locks had never a key—
What kind of box, then, could it be?

And now, Carroll's poetic solution:

As curly-headed Jemmy was sleeping in bed,
His brother John gave him a blow to the head;
James opened his eyelids, and spying his brother,
Doubled his fist, and gave him another.
This kind of box then is not so rare;
The lids are the eyelids, the locks are the hair,
And so every schoolboy can tell to his cost,
The key to the tangles is constantly lost.

A seemingly non-answerable Carroll conundrum is asked by the Mad Hatter: "Why is a raven like a writing-desk?" Carroll first posed it without a solution and in later editions appended this solution: "Because it can produce a few notes, though they are very flat; and it is never put with the wrong end in front." In early revisions he spelled "never" as "nevar," which is raven "with the wrong end in front."

Figure 8.2 Photograph of Raymond Smullyan (1919–2017). Photo credit: The Raymond Smullyan Society.

The concert pianist, philosopher, magician, mathematical logician, and master puzzle-maker Raymond Smullyan (Fig. 8.2) claims that the following syllogism is valid.[11]

1. Everyone loves my baby.
2. My baby loves only me.
∴ I am my own baby.

But is it really valid? Let us ruminate on it for a while...

Here's one explanation, slightly different from Smullyan's. By premise 2, my baby loves a single person—me. And by premise 1, my baby, who is a person (and thus included in "everyone"), loves my baby. So, my baby loves only me by 2, and, by 1, loves my baby. That must mean that I'm my own baby. Convinced?

Before evaluating another of Smullyan's unusual syllogisms,[11] first recall that in Shakespeare's great play *Othello*, Iago hates Othello. Now:

1. Everyone loves a lover.
2. Romeo loves Juliet.
∴ Iago loves Othello.

Exercise 17: Why is the syllogism valid, or why is it not?

Chapter 9
Archimedes

Figure 9.1 Portrait of Archimedes (ca. 287–212 BCE). Woodcut from *Les Vrais Pourtraits et Vies des Hommes Illustrés Grecz, Latins, et Payens* by André Thevet (1584).

The story of Archimedes, the golden crown, and "eureka" is well known. But it's interesting enough to bear repeating.

King Hiero II of Syracuse, a city on the eastern coast of ancient Sicily, commissioned a goldsmith to make him a crown of pure gold in the shape of a wreath. After the crown had been made, Hiero began hearing rumors that the goldsmith may have cheated him by mixing with the gold some less costly silver. King Hiero consulted his cousin Archimedes (Fig. 9.1), who had already distinguished himself in mathematics, mechanics, and physics, though he was then only 22 years of age.

Could Archimedes determine whether the crown was part silver or pure gold? There was no question of melting it down because Hiero wanted the wreath to remain intact.

Archimedes was initially stumped. One day, while pondering the problem, he stepped into a public bath. Noticing that immersing his body into the bath displaced a certain quantity of water, Archimedes was suddenly inspired.

"Eureka, eureka!"—"I've got it, I've got it!"—he shouted, running through the streets of Syracuse naked (for he'd forgotten to put on his clothes).

Archimedes realized that he could test the purity of the crown by immersing it in water and measuring how much water was displaced. First, he took a bar of pure gold that exactly balanced the weight of the crown. If the crown were pure gold, it would displace the same volume of water as the bar of gold.

But the crown displaced more water than an equal weight of gold. Why? Because gold weighs more than silver, it stands to reason that a crown mixed with silver (or another metal lighter than gold) would require more material to reach the same weight as one composed only of gold. History does not tell us what happened to the goldsmith. It's best left to our imaginations.

Questions have been raised about the accuracy of Archimedes' method. Also, it would have been very difficult for him to have found a quantity of gold that weighed exactly the same as the crown. Still, we can imagine that Archimedes' method, however inaccurate, was precise enough to reveal the goldsmith's deception.

Eighteen centuries later, another 22-year-old derived a more accurate way to evaluate the purity of Hiero's golden crown. Galileo (Fig. 9.2), now known as the father of modern science, combined Archimedes' method of submerging the crown in water to determine whether it was pure gold, and Archimedes' use of the lever.

Figure 9.2 Portrait of Galileo Galilei (1564–1642). Image credit: Italian painter, 17th century, after Justus Suttermans / Wellcome Images / licensed under CC BY 4.0.

First, Galileo placed a quantity of pure gold on one side of a lever and an alloy of gold and silver on the other side. Then he adjusted the pivot point of the lever so that they balanced exactly. Then, when he immersed the entire balance in water, he took note that the end of the lever holding the alloy tilted upwards. From this, he determined how much to adjust the pivot point to account for the alloy weighing less than the quantity of pure gold.

Ratio of the Circumference of a Circle to Its Diameter

Another accomplishment of Archimedes was to arrive at an extremely accurate approximation of the circumference of a circle, given its diameter. That is, he found a close approximation to π in the formula where d is the circle's diameter, and π multiplied by d equals its circumference. It has been said that the ancient Egyptians considered the ratio of the circumference of a circle to its diameter to be 3⅓, which is not very accurate. 3⅓ ≈ 3.33333. (It has been claimed that in the Bible the ratio is just 3 but, as you may imagine, this is controversial.)

The way Archimedes arrived at his calculation was to inscribe and circumscribe regular polygons both inside and around the circumference of a circle, and then take their average. Since he used 96-sided polygons, that method provided an extremely close estimate. Figure 9.3 shows a circle with two hexagons circumscribed and inscribed about a circle.

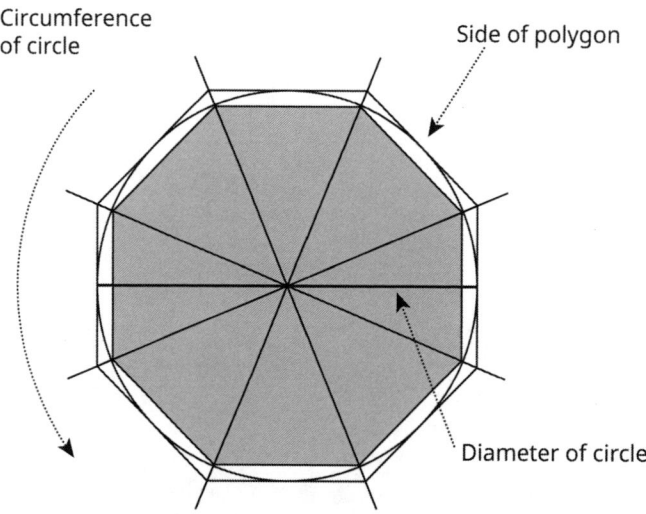

Figure 9.3   Illustration of Archimedes' method of approximating π. Image credit: Ernesto Mora.

Archimedes' approximation was 3 10/71 $<$ π $<$ 3 1/7. In decimal notation, these fractions are approximately 3.140845 and 3.1428571, respectively. The midpoint of these two numbers is 3.141858, which is 99.9% accurate.

Using an extension of the method above, often referred to as the "Method of Exhaustion," Archimedes is said to have invented the calculus of Leibniz and Newton almost 2000 years before them. This claim seems hard to support, as any method using only finitely many polygons doesn't invoke a limit, no matter how many polygons Archimedes used.

A bust of Archimedes is featured on the Fields Medal in mathematics (Fig. 9.4), the highest honor given to young mathematicians for outstanding work. Often referred to as the mathematicians' Nobel Prize, it is presented every four years to a small number of young mathematicians responsible for mathematical breakthroughs.

Figure 9.4 Photograph of a Fields Medal featuring Archimedes. Photo credit: Stefan Zachow / Wikimedia Commons.

On the face of the medal, written in uppercase Greek letters, is the word ΑΡΧΙΜΗΔΟΥΣ, meaning "of Archimedes," and a quote in Latin attributed to him: *Transire suum pectus mundoque potiri*—"Rise above oneself and grasp the world." The date, written in Roman numerals, is mistakenly inscribed with MCNXXXIII rather than MCMXXXIII.

Thanks to Archimedes, we know how to find the approximate value of π, the ratio of the circumference of a circle to its diameter. But can we calculate the circumference of the Earth?

That was left for Eratosthenes to do.

But before we move on to that solution, let's take a quick look at another problem Archimedes hurled at Eratosthenes, and at others. The problem is nifty, but solving it is incredibly difficult. It's so difficult that it wasn't solved until approximately 1880. The smallest solution is about $7.76 \cdot 10^{206,544}$, which is a much, much larger number than the number of atoms in the known universe, estimated at between 10^{78} and 10^{82} atoms. The actual number is too large to be computed by humans.

Here's the problem posed by Archimedes. How far can you get before being stumped?

> If thou art diligent and wise, O stranger, compute the number of cattle of the Sun, who once upon a time grazed on the fields of the Thrinacian isle of Sicily, divided into four herds of different colors,

one milk white, another a glossy black, a third yellow and the last dappled. In each herd were bulls, mighty in number according to these proportions: Understand, stranger, that the white bulls were equal to a half and a third of the black together with the whole of the yellow, while the black were equal to the fourth part of the dappled and a fifth, together with, once more, the whole of the yellow. Observe further that the remaining bulls, the dappled, were equal to a sixth part of the white and a seventh, together with all of the yellow. These were the proportions of the cows: The white were precisely equal to the third part and a fourth of the whole herd of the black; while the black were equal to the fourth part once more of the dappled and with it a fifth part, when all, including the bulls, went to pasture together. Now the dappled in four parts were equal in number to a fifth part and a sixth of the yellow herd. Finally the yellow were in number equal to a sixth part and a seventh of the white herd. If thou canst accurately tell, O stranger, the number of cattle of the Sun, giving separately the number of well-fed bulls and again the number of females according to each colour, thou wouldst not be called unskilled or ignorant of numbers, but not yet shalt thou be numbered among the wise.

But come, understand also all these conditions regarding the cattle of the Sun. When the white bulls mingled their number with the black, they stood firm, equal in depth and breadth, and the plains of Thrinacia, stretching far in all ways, were filled with their multitude. Again, when the yellow and the dappled bulls were gathered into one herd they stood in such a manner that their number, beginning from one, grew slowly greater till it completed a triangular figure, there being no bulls of other colors in their midst nor none of them lacking. If thou art able, O stranger, to find out all these things and gather them together in your mind, giving all the relations, thou shalt depart crowned with glory and knowing that thou hast been adjudged perfect in this species of wisdom.

Chapter 10
Eratosthenes

The Circumference of the Earth

Eratosthenes arrived at an ingenious method for calculating the circumference of the Earth. The date was about 240 BCE. If you're like me, you were taught in history class that Christopher Columbus was the first to measure the Earth's circumference (around 1477). Columbus' measurement was actually far below the value accepted today. By contrast, Eratosthenes' measurement, more than 1700 years before him, was extremely accurate.

Figure 10.1 Portrait of Eratosthenes (ca. 276–194 BCE). Copper engraving by Gottlieb Friedrich Riedel from *Gallerie der alten Griechen und Römer*, pp. 236–237, Universitätsbibliothek Heidelberg (1801).

Here's how Eratosthenes did it. First, the Greeks knew—possibly as long ago as 600 BCE—that the Earth was round because they could see that departing ships dropped out of sight over the horizon before their masts did. Also, Eratosthenes knew that in Egypt there was a deep well in the city of Syene (now Aswan). At noon during the summer solstice, around June 21, the Sun's rays were directly overhead at Syene, as evidenced by the rays going precisely down the well, missing all sides and hitting the water at the bottom.

It was assumed that the Sun is so far from the Earth that its rays are parallel to the Earth. Virtually directly north of Syene was Alexandria. The distance between the two cities was measured as 5000 "stadia" or stades. Eratosthenes had a pole (a "gnomon") erected in Alexandria that pointed directly overhead —probably a vertical sundial that pointed straight up. At the moment when the Sun's rays hit the exact bottom of the well in Syene, the angle between the pole (in Alexandria) and its shadow was approximately 7.2° or 1/50th of the circumference of a circle. (Eratosthenes' measurement of the shadow was actually only 7.12°, but he rounded up to 7.2 in order to divide 360 evenly.)

Consult Fig. 10.2.

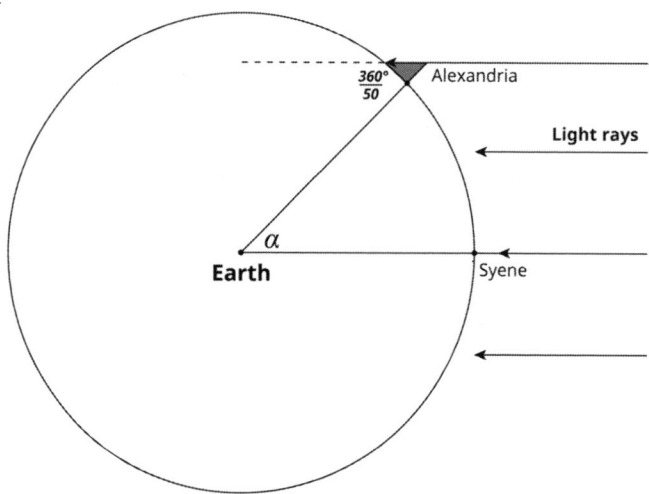

Figure 10.2   Illustration of Eratosthenes' method of calculating the circumference of the Earth. Image credit: Ernesto Mora.

As we covered in Chapter 6, in geometry, alternate interior angles between parallel lines are equal. 7.2°, as shown above, is 1/50th of the circumference of a circle. Thus, the circumference of the Earth is 50 times the number of stadia (stades) between Syene and Alexandria, which is approximately 500 miles. Multiplying 500 miles by 50 equals 25,000 miles, which was Eratosthenes' calculation of the circumference of the Earth. This figure is less than 1% greater than the actual circumference of the Earth, which is 24,901 miles.

The Sieve of Eratosthenes

It would be a shame to mention Eratosthenes' method for calculating the circumference of the Earth without looking at his well-known and often used sieve for obtaining prime numbers. Eratosthenes' "sieve" to extract primes from the positive integers starting with 2 (1 is not a prime number; 2 is the first) is extremely efficient. It is easily programmed in every computer programming language, and usually all primes up to millions of numbers can be found in a split second. How does it work?

Take one million numbers, for example, and list them in order, starting with 2: 2, 3, ..., 1,000,000. Print the first number in the list, which is 2. Then cross out all multiples of 2 in the list. Print the first number in the remaining list, which is 3. Again, divide all remaining numbers by 3 and cross them out. The next remaining number is 5 (since 4 has been eliminated, as it's divisible by 2). So, cross out all subsequent numbers divisible by 5. Then go to 7 (6 is a multiple of both 2 and 3 and has thus been crossed out). Then, just keep on going in the same way until you've run through the one million numbers. This sieve method filters out all non-primes.

As a simple example, we will take all numbers starting with 2 and going up only to 30, showing the prime numbers that have not been filtered out by Eratosthenes' sieve. Programming this in some languages requires a loop within a loop. The first loop continues to the limit number. The inside loop crosses out multiples of primes obtained earlier. In the simple example below, the limit equals 30.

The numbers below in bold are primes; the rest are composite numbers:

2, **3**, 4, **5**, 6, **7**, 8, 9, 10, **11**, 12, **13**, 14, 15, 16, **17**, 18, **19**, 20, 21, 22, **23**, 24, 25, 26, 27, 28, **29**, 30.

As you can see, only 2, 3, 5, 7, 11, 13, 17, 19, 23, and 29 remain; they're all primes!

A simplification should probably be mentioned. When checking for primes up to some number n, it is necessary to check only numbers not exceeding the square root of n. This may not be obvious, but you should be able to see it for yourself. If a number n is not prime, it can be factored into two factors, call them a and b. Factors a and b can't *both* be bigger than the square root of n, or else multiplying them together would give you a value higher than n. So, in any factorization of n, at least one of the factors must be smaller than the square root of n. If we can't find any factors less than or equal to the square root of n, then n must be prime.

Hence, in our example of finding primes up to one million, it is necessary to check primes only up to one thousand, not the full million. This saves a great deal of time (and storage capacity of a computer). For instance, to find all primes up to 100, only 2, 3, 5, and 7 need to be used, since the next prime on the list after 7 is 11, which is greater than 10, the square root of 100. These four numbers, 2, 3, 5, and 7, are the only ones needed to find all primes up to 120; then 11 is needed for 121.

Here's an example of pseudocode for an overly simple computer program that prints all primes from 2 to some limit (omitting details—for instance, "Erase all multiples of i" is a loop).

```
Sieve (numbers up to: Limit)
    For i from 2 to √Limit
        Print i
            Erase all multiples of i
```

The sieve of Eratosthenes is so nifty that it deserves to be listed in Erdős' "Book."

Chapter 11
More Ancient Wisdom: Theodorus, Plato, Hypatia, and Thales

Figure 11.1 Portrait of Theodorus (ca. 465–398 BCE). Image credit: Österreichische Nationalbibliothek, Wien PORT_00149568_02.

Theodorus is said by Plato in his dialogue *Theaetetus* to have proven that the square root of each non–perfect-square number besides 2, up to 17, is irrational. A positive irrational number is a positive integer that cannot be expressed as a fraction (a rational number) or a terminating decimal (which later will be shown to be the same as a fraction). The first few perfect squares up to 17 are 1, 4, 9, and 16 (respectively equal to 1^2, 2^2, 3^2, and 4^2). All the rest, 2, 3, 5, 6, 7, 8, 10, 11, 12, 13, 14, 15, and 17, have irrational square roots. Theodorus is said to have proved these numbers irrational by use of what has come to be called "the spiral of Theodorus" (Fig. 11.2).

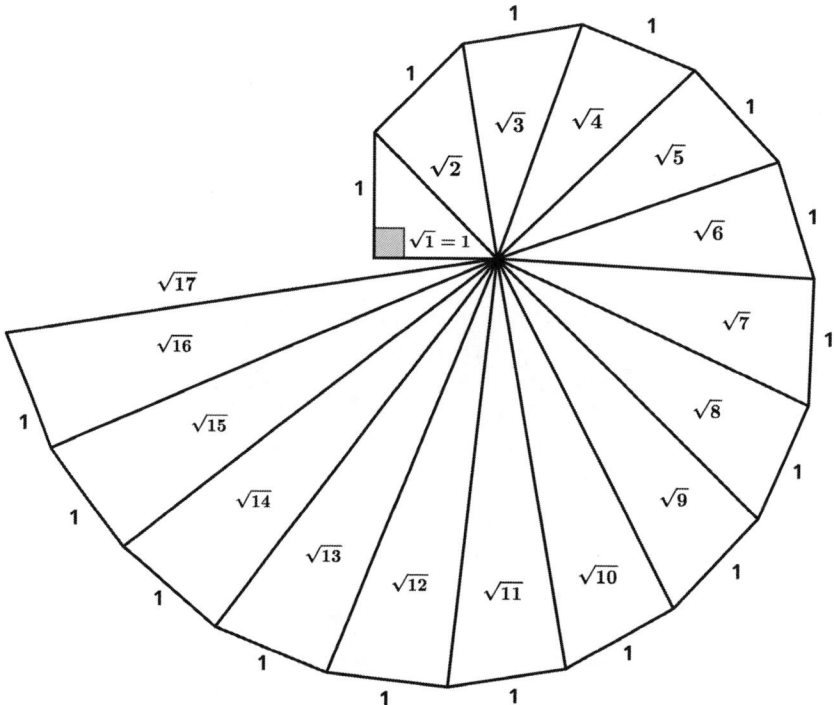

Figure 11.2 Illustration of the spiral of Theodorus. Image credit: Ernesto Mora.

The attractiveness of Theodorus' spiral is the primary reason for displaying it, since no information exists to explain how it was used in a proof. The spiral is obtained by starting with a triangle whose two legs are each 1, as shown above, with the darkened square to indicate a right angle. The hypotenuse of this smallest triangle is $\sqrt{2}$. The adjacent triangle is created by using as the length of one leg the earlier hypotenuse, $\sqrt{2}$. The other leg of this triangle is length 1, formed by drawing a line perpendicular to that hypotenuse. Thus, the length of its hypotenuse is $\sqrt{3}$, the square root of $(\sqrt{2})^2 + 1^2$. Similarly, the next triangle is formed by taking as the length of one leg the earlier hypotenuse ($=\sqrt{3}$), and again taking 1 to be the length of the second leg, drawn perpendicular to that hypotenuse. This makes the length of the new hypotenuse the square root of $(\sqrt{3})^2 + 1^2$, which equals $\sqrt{4}$, and $\sqrt{4} = 2$. And so on.

Since the spiral yields both irrational and rational numbers, the question arises how Theodorus distinguished one from the other. The answer is unknown. Nonetheless, this spiral is frequently cited as having been used to separate irrational numbers. Many hypotheses have been advanced as to how it should be applied, but nothing is certain. At any rate, this spiral, no matter how it was used, establishes that in some sense Theodorus was an interesting early mathematician.

Figure 11.3 Line engraving of a bust of Plato (427–347 BCE). Engraving by L. Vorsterman after Sir P. P. Rubens / Wellcome Images / licensed under CC BY 4.0.

In *Meno*, a dialogue of Plato that explores the nature of virtue, Socrates poses a mathematical puzzle to a slave boy. First, Socrates draws a square in the sand whose sides are each of length 2, making the area of the square equal to 4. He then asks the boy to modify the figure to obtain a square whose area is double that area, making the resultant area equal to 8.

Socrates' square is shown in Fig. 11.4.

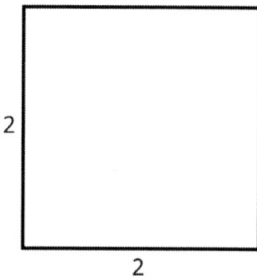

Figure 11.4 Illustration of Socrates' square (Meno's slave). Image credit: Ernesto Mora.

The boy's answer is to double the length of each side, which he believes will produce a square with double the area of the above square. Socrates follows the boy's directions by drawing the squares in Fig. 11.5. He then asks the boy to tell him the area of the large square.

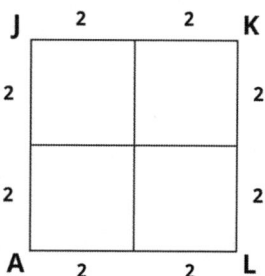

Figure 11.5 Illustration of Socrates' square (Meno's slave) quadrupled. Image: Ernesto Mora.

The boy sees that the area of the large square is 16—four rather than two times that of the single square. Socrates reminds him that the question was to modify the area of the 2×2 square so that the resultant area is only two times the area of the small square. That is, it should be 8, not 16. The slave boy is then puzzled how to proceed.

Exercise 18: Modify the square produced by Socrates' pupil such that you get a square whose area is 8, half the area of the one arrived at by his method.

HYPATIA

Figure 11.6 Portrait of Hypatia (ca. 370–415). Illustration by Jules Maurice Gaspard from *Little Journeys to the Homes of Great Teachers: Hypatia* by E. Hubbard (1908).

One of the very few ancient women mathematicians of whom anything is known was the brilliant and beautiful Hypatia of Alexandria. Widely known for her expertise in both mathematics and philosophy, as well as for her generosity in teaching these subjects, she was attacked in the year 415 by a mob of Christian monks who felt that her teachings threatened the rise of Christianity. Accounts differ as to how she was killed, but what's agreed on is that she was brutally slain. Consequently, Hypatia has come to be thought of as a martyr, whose death marks the end of the secular, classical era. Unfortunately, none of her writings has survived.

106. Thales.

Figure 11.7 Drawing of a bust of Thales (ca. 624–548 BCE). [Line engraving by Wilhelm Meyer from *Illustreret Verdenshistoria* by P. Laessoe-Muller and J. Ostrup (1927).

According to Aristotle and many other ancient Greeks, Thales was the very first scientific Greek philosopher. He observed that life emanates from moist substances and postulated water as the fundamental substance from which everything in the universe emerges. This is not so far-fetched when one considers that water exists in three forms: ice, liquid, and mist—a solid, liquid, and gas. Before him were poets who spun mythological tales about gods and goddesses originating all of life.

Thales purportedly predicted the solar eclipse of 585 BCE and also cornered the olive oil market by buying up rights to use all available olive presses ahead of a bonanza harvest. Besides the wealth accrued, Thales supposedly wished to show others that philosophy was practical. However, it was also said of Thales that he was very absent-minded, and there persisted a rumor that while stargazing one night, he tumbled headlong into a ditch.

Thales is credited for originating the first strictly mathematical proofs. He was said to have learned some geometric principles from the Egyptians, who used empirical rather than theoretical methods to arrive at their results. One of Thales' proofs is labeled "Thales' theorem":

If A, B, and C are points on a circle, and AB is a diameter, then ∠ACB is a right angle (Fig. 11.8).

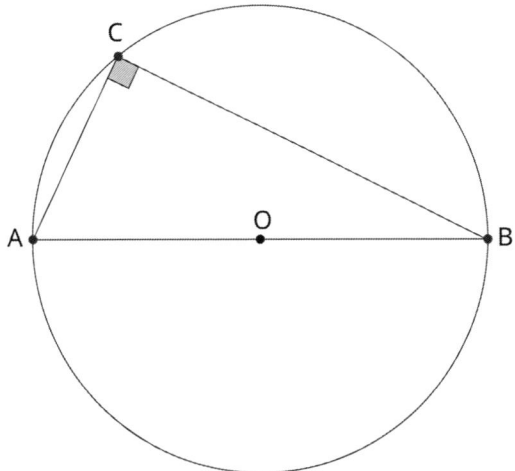

Figure 11.8　Diagram of Thales' theorem. Image credit: Ernesto Mora.

For proof, consult Fig. 11.9. Connect OC. Then, OC and OA are equal because they are both radii, which makes ∠OCA = ∠OAC (both labeled x). And ∠OCB = ∠OBC (both labeled y) for the same reason. Thus, summing the angles in triangle ABC, we have $2x + 2y = 180°$. Rewriting this as $2(x + y) = 180°$, the ∠x + ∠y, which is the same as ∠ACB, is 90°, a right angle. Which is what we wished to show.

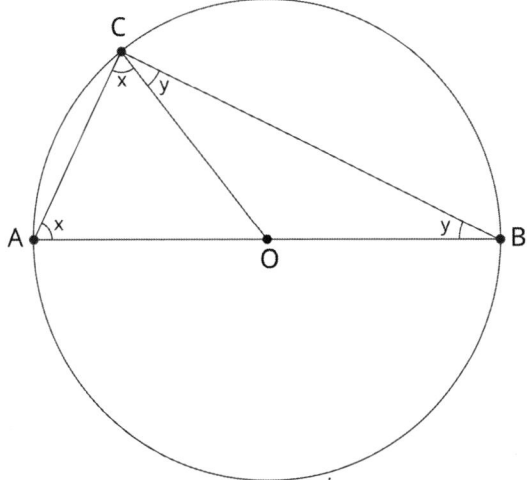

Figure 11.9　Diagram of a proof of Thales' theorem. Image credit: Ernesto Mora.

Changing Repeating Decimals into Fractions

This concept is elementary, but still nifty.

Take the repeating decimal 0.636363.... Since the decimal repeats, we know that there must be an equivalent fraction. The problem for you is to find the fraction equal to this decimal. Do you know an easy way to find it?

Since 0.636363... repeats after two numbers, we want to multiply it by 100 (you'll soon see why), and then we'll subtract the rational decimal in question:

$$\begin{array}{r} 100x = 63.636363\ldots \\ -\,1x = 0.636363\ldots \\ \hline 99x = 63 \end{array}$$

When we divide both sides of the equal sign, we get

$$x = 63/99 = 7/11$$

The reason for multiplying by 100 is that 0.63 has two places before repeating, and multiplying by 100 moves the top number two places to the left. When the top number was moved over two places, it was easy to subtract the bottom number from it.

If the repeating decimal has 6 numbers before repeating, we move over 6 digits and subtract again, which cancels out the decimal portion. Here's an example that repeats after 6 numbers: 0.076923076923....

$$\begin{array}{r} 1{,}000{,}000x = 76{,}923.076923\ldots \\ -\phantom{000{,}00}1x = \phantom{76{,}923.}0.076923\ldots \\ \hline 999{,}999x = 76{,}923 \\ x = 76{,}923/999{,}999 = 1/13 \end{array}$$

Of course, this is neither a spectacular nor difficult result, but I hope you agree that it's clever, especially when the long fractions can be reduced, as in the last case, to 1/13.

What is of significant mathematical interest is that this method shows how to convert any terminating or repeating decimal into a fraction, which establishes that the two are interchangeable.

Exercise 19: Convert 0.142857142857... to a fraction in lowest terms. Try out some other repeated decimals on your own and see whether you can find a few more tidy ones.

Chapter 12
Probabilities

How to be Smarter than (Some) Mathematicians

Figure 12.1 Photograph of Marilyn vos Savant. Photo credit: allthatsinteresting.com / Wikimedia Commons.

A most interesting and popular case of counterintuitive probability was super-genius Marilyn vos Savant's unusual solution to what seemed a straightforward problem. Vos Savant's result fooled numerous people, many of whom were mathematicians. Here is the problem:[12]

> Suppose you're on a game show, and you're given the choice of three doors. Behind one door is a car, behind the two other doors are goats. You pick a door, say #1, and the host, who knows what's behind the doors, opens a door with a goat, say #3. He says to you, "Do you want to pick door #2?" Is it to your advantage to switch your choice of doors?
>
> —Submitted by Craig F. Whitaker, Columbia, Maryland

This was the question posed of Marilyn vos Savant in *Parade* magazine. It's often called the "Monty Hall problem," named after the original host of the television game show *Let's Make a Deal*.

Her answer: Yes, you should switch. Switching doors, she wrote, improves the probability of getting the car from 1/3 to 2/3, i.e., twice the probability for switching versus not switching.

It should be mentioned that Marilyn vos Savant was listed years ago in *The Guinness Book of World Records* as having the highest IQ ever tested: 228, recorded on the Stanford–Binet exam when vos Savant was 10. For several reasons her IQ has been challenged. Many high IQers claim to be smarter than vos Savant. Some persons just believe an IQ test of a child cannot be correlated with the score on an adult test. And there now exist many, many other IQ tests, each one claimed to be more accurate than the others for specific talents. Some tests have been specifically crafted to determine super-duper genius-hood. For instance, the Giga Test can supposedly find "one in a billion"—i.e., anyone who can outscore 0.999999999 of the adult population in intelligence. It has been estimated that there are 7 or 8 such Giga giant IQers living today (though there are fewer than 8 billion adults alive!). Despite this confusion of claims and results, vos Savant's name still appears as one of the top three geniuses on multiple lists. And many of those on the list have never even been tested. Their IQs are simply estimated, based on claims of their amazing precociousness when children.

I cannot refrain from this addendum: two of the top Giga-geniuses challenging vos Savant's IQ supremacy are both "iron freaks" (avid body builders) who've worked in bars as bouncers. Perhaps more interesting, one of the highest IQs ever recorded belonged to the convicted murderer Nathan Leopold. In Leopold's autobiography, *Life Plus 99 Years*,[13] he describes taking one IQ test after another during 23 years of incarceration, just to see to what degree his intelligence declined in prison. He claimed that his IQ remained steady during his many years of confinement.

Vos Savant was flooded with thousands of responses to the game show problem. Among over 10,000 respondents, 75% of them said she was wrong. Many said she was an idiot. This included a good number of mathematics PhDs. Here are just a few of their letters[12] (taken directly from vos Savant, who was forced to devote three subsequent columns to explaining why her reasoning was correct):

> Since you seem to enjoy coming straight to the point, I'll do the same. You blew it! Let me explain. If one door is shown to be a loser, that information changes the probability of either remaining choice, neither of

which has any reason to be more likely, to 1/2. As a professional mathematician, I'm very concerned with the general public's lack of mathematical skills. Please help by confessing your error and in the future being more careful.
—Robert Sachs, PhD, George Mason University

You blew it, and you blew it big! Since you seem to have difficulty grasping the basic principle at work here, I'll explain. After the host reveals a goat, you now have a one-in-two chance of being correct. Whether you change your selection or not, the odds are the same. There is enough mathematical illiteracy in this country, and we don't need the world's highest IQ propagating more. Shame!
—Scott Smith, PhD, University of Florida

Your answer to the question is in error. But if it is any consolation, many of my academic colleagues have also been stumped by this problem.
—Barry Pasternack, PhD, California Faculty Association

I have been a faithful reader of your column, and I have not, until now, had any reason to doubt you. However, in this matter (for which I do have expertise), your answer is clearly at odds with the truth.
—James Rauff, PhD, Millikin University

May I suggest that you obtain and refer to a standard textbook on probability before you try to answer a question of this type again?
—Charles Reid, PhD, University of Florida

I am sure you will receive many letters on this topic from high school and college students. Perhaps you should keep a few addresses for help with future columns.
—W. Robert Smith, PhD, Georgia State University

You are utterly incorrect about the game show question, and I hope this controversy will call some public attention to the serious national crisis in mathematical education. If you can admit your error, you will have contributed constructively towards the solution of a deplorable situation. How many irate mathematicians are needed to get you to change your mind?
—E. Ray Bobo, PhD, Georgetown University

I am in shock that after being corrected by at least three mathematicians, you still do not see your mistake.
—Kent Ford, Dickinson State University

Maybe women look at math problems differently than men.
—Don Edwards, Sunriver, Oregon

> You made a mistake, but look at the positive side. If all those PhDs were wrong, the country would be in some very serious trouble.
> —Everett Harman, PhD, U.S. Army Research Institute

And last, but not least:

> You are the goat!
> —Glenn Catkins, Western State College

The arrogant self-assurance of these math professors' published remarks would haunt them long into their careers once it became clear that vos Savant was right and they were wrong. Imagine how other professors and students would no doubt have mocked them as they had mocked Marilyn vos Savant.

There are multiple correct ways of analyzing the Monty Hall problem, all of them arriving at the conclusion that vos Savant was right. The correct answer to the probability of switching is 2/3 rather than the (initial) probability of 1/3. This may seem counterintuitive, so let's try to simplify the problem.

There are two steps involved: the first is the initial choice and the second is the decision to stay with that choice or switch choices. Let's look at both cases. In step 2, staying with the initial decision, which was correct one time in three, yields the same probability of success: 1/3. Switching from the initial decision reverses the probability from being wrong in step 1 to being right in step 2. Since the probability of being wrong in step 1 is 2/3, then the probability of being right after switching becomes 2/3. Thus, vos Savant is correct: you should switch.

Since the above explanation may still be unintuitive, let's try another way to view the problem. This is one of the ways vos Savant suggests. She writes: "The benefits of switching are readily proven by playing through the six games that exhaust all possibilities. For the first three games you choose number 1 and 'stay' each time. For the second three games you choose number 1 and 'switch' each time and the game show host always opens a loser. Here are the results."[12]

I'll underline the choice, which in this scenario is always 1. Note that it doesn't matter which your pick is. Here it's 1, but it could have been 2 or 3.

1. CAR Goat	Goat	I stay and win.	
2. Goat CAR	Goat	I stay and lose.	→ WIN 1/3
3. Goat Goat	CAR	I stay and lose.	

1. <u>CAR</u> Goat   Goat   I switch and lose.
2. <u>Goat</u> CAR   Goat   I switch and win.   → WIN 2/3
3. <u>Goat</u> Goat   CAR   I switch and win.

The ingenious cartoonist Larry Gonick solved the problem this way: if the contestant chooses wrong, the probability of winning by switching is 1 (since the switch would reveal the car). Therefore, the contestant is wrong 2/3 of the time. The probability of winning by switching would then be the product $1(2/3) = 2/3$.

To generalize somewhat, suppose there were four doors—three goats and one car. Applying the above method, if the contestant chooses wrong, the probability of winning by switching is 1/2 (because two goat doors have been eliminated). The contestant is wrong 3/4 of the time. The probability of winning by switching is therefore $(1/2)(3/4) = 3/8$, which is much better than not switching, which wins only once in four. This method seems to generalize to any number of doors, where the number of cars plus the number of goats equals the number of doors.

Incidentally, many persons who quarreled with vos Savant's answer argued that her statement of the problem was ambiguous between supposing that the host of the game show knew which door had the car and supposing he had not known. They argued that her analysis wouldn't work had he not known. But this is not correct, for if he happened to pick the door with the car, he merely could eliminate that pick and reset the situation again until he picked a goat door.

In case you're still not convinced, set up a way to test the stay–switch technique experimentally. Or, if you're enterprising, write a computer program to test this hypothesis. The program will reveal, supposing you're still in doubt, that switching is superior (twice as good as staying). Unfortunately, I know a recalcitrant doubter who wrote just such a computer program. Sure enough, his program established that switching won 2/3 of the time, whereas staying with the original guess won only 1/3 of the time. Unfortunately, his refusal to learn something dominated the result of his own computer program. Grumpily, he said that his program bore out Marilyn vos Savant's analysis, but he still didn't believe it. (I know there's a moral here, but I'm not sure what it is!)

More Probabilities

Here's a well-known but nevertheless neat problem: What is the probability that any two of 25 persons taken at random will have the same birthday?

Whatever the probability is, it must be quite small, don't you think? Take a wild guess, supposing you don't already know the answer. Did you guess a small percentage, like 5%? Or even lower? What did you guess?

Well, surprisingly, the probability of any two of 25 random people—those in an average size classroom, for example—will have the same birthday is well over 50%. It turns out that when about 23 people are asked their birthday, the odds are slightly over 50-50 that two of them will have the same one.

To explain this will take us to the heart of probability theory, which some of you would surely like to avoid. Well, we'll just drop in on probability theory to snag some interesting notions. Hopefully, these snagged items will add up to an explanation.

First, we should explain why any two of 23 persons picked at random are just as likely to share a birthday as not. We don't care about the year they were born or anything else, just the month and date of their birthday. Oh, I almost forgot, leap year birthdays are not included simply because one date, February 29, comes up every four years (often, but not always).

Take a look at this number: 364/365.

What do you make of it? It's the probability that two persons will have different birthdays. Person 1 has 365 days for a birthday, but Person 2 has only 364 choices of a different birthday after one possible birthday has been used up by Person 1. Similarly, Person 3 will have left only 363 different birthday possibilities, since both 364 and 365 have been eliminated. Moving right along, we arrive at:

$$364/365 \quad + \quad 363/365 \quad + \quad 362/365 \quad + \quad 361/365 \quad + \ldots + \quad 343/365$$
two different   three different   four different   five different   23 different

Recall that the probability that some event does not occur plus the probability that the event does occur must add up to one. So, one minus the probability of an event not occurring equals the probability of that event occurring. In the present case, if $364/365 + 363/365 + 362/365 + 361/365 + \ldots + 343/365$ represents all possibilities that no two of 23 persons have birthdays on the same day of the year, then $1 - (364/365 + 363/365 + 362/365 + 361/365 + \ldots + 343/365)$ must equal the probability of the event actually occurring, which turns out to be slightly above 50%—actually, 50.73%. So, if you're in a room with more than 30 people in it, it's a good bet that two of them share a birthday since that probability is over 70%.

Counterintuitive, isn't it? That makes it nifty in my opinion. I hope in yours as well.

We might as well fiddle with probabilities now that we've started. Here's something a gamer, trying to win your money rolling dice, might try. Say he wants to bet you that if the numbers on two dice add up to 2, 3, 4, or 5, you win. If they add up to 6, 7, or 8, he wins. If they yield 9, 10, 11, or 12, neither of you wins. At first glance, it seems you have the advantage. You have four possibilities to his mere three.

Let's calculate the probabilities. First of all, there are 36 total ways that two dice can show up—six possibilities for each die ("die" is the singular of "dice"; the dots on a die are called "pips"). To help us calculate this correctly, let's suppose that one die has red pips and the other has green.

How many ways can a 2 show up? Only one: one red pip and one green pip.

For a 3 to appear, there can be two red pips and one green pip, or two green pips and one red pip. That's two possibilities.

For the number 4, there can be two red pips and two green pips, or three red pips and one green pip, or three green pips and one red pip. The result is that there are three possibilities for two dice adding up to four.

These explanations are becoming so wordy that it seems better to create a chart.

Sum of Pips	# of Possibilities
2	1
3	2
4	3
5	4
6	5
7	6
8	5
9	4
10	3
11	2
12	1

Note that the total number of possibilities equals 36. The chart also shows that the sum of the possibilities for getting a total of 2, 3, 4, or 5 is 10. The sum of possibilities for getting a 6, 7, or 8 is 16. The gamer is gaming you, making you think you have an advantage because you have four numbers compared to his three. But in truth, he'll win 16 times to your 10, or eight to your five, which is quite an advantage!

How is it that a sum of seven pips can show up six different ways—more than any other sum? First, look at the sum alone: $7 = 6 + 1, 5 + 2,$ and $4 + 3$. That's only three possibilities. But recall that one die has red pips and one die has green pips. We need to double our possibilities to account for this. It could be six red pips and one green, or six green pips and one red. And so on.

It doesn't matter that we colored the pips on our dice differently. They could all be red, for example. But it's then more difficult to explain how there are really two possibilities. We'd have to refer to Die 1 and Die 2, which is less clear than "red" or "green."

Exercise 20: One more probability puzzle. Suppose you have three cards. One of them is red on both sides, one is green on both sides, and the third is red on one side and green on the other. If you grab a card randomly and one side is red, what are the chances that the other side is red as well?

Chapter 13
Paradoxes

Before we begin, we need to define what we take a paradox to be. There are multiple definitions. A paradox, by our definition, is a statement that, despite apparently sound reasoning from acceptable premises, leads to a conclusion that seems impossible.

The Liar Paradox

The first instance of a strict version of the liar paradox comes from Eubulides of Miletus, who lived during the fourth century BCE. Famous for several paradoxes, Eubulides reportedly asked, "A man says that he is lying. Is what he says true or false?"

Why should we care about such a riddle? Søren Kierkegaard (1813–1855) wrote that "one must not think ill of the paradox, for the paradox is the passion of thought, and the thinker without the paradox is like the lover without passion: a mediocre fellow."[14] So, upon advisement, we find paradoxes of interest.

First, let's consider a "partial" liar paradox.

Figure 13.1 Illustration of Epimenides (sixth century BCE). Illustration by Michael Wolgemut from the *Nuremberg Chronicle* (1493) / Library of Congress.

Epimenides did not write "I am lying"; he wrote instead that all Cretans are liars. And, as a matter of fact, he was born in Crete, making him a Cretan. The source of this paradox is a poem Epimenides wrote that claimed Zeus was immortal:

> They fashioned a tomb for thee, O holy and high one
> The Cretans, always liars, evil beasts, idle bellies!
> But thou art not dead: thou livest and abidest forever,
> For in thee we live and move and have our being.

Note that the above poem asserts of Cretans that they're "always liars." This raises a question: Was Epimenides, the author of the poem and a Cretan, contradicting himself? Let's look at the statement as a syllogism in modern logical notation.

1. $\forall x(Cx \rightarrow Lx)$ All Cretans are always liars (for all x, if x is a Cretan, then x is always a liar).
2. Ce Epimenides is a Cretan.
3. $\therefore Le$ Therefore, Epimenides is always a liar.

But if Epimenides lies when he says that all Cretans are always liars, then at least one Cretan tells the truth sometimes (but not Epimenides, who we've seen is a liar). We can symbolize this transformation as follows:

1. $\neg\forall x(Cx \rightarrow Lx)$ It's not the case that all Cretans are always liars.
2. $\exists x(Cx \wedge \neg Lx)$ Some Cretan exists who is not always a liar—(s)he sometimes tells the truth.
3. $Ck \wedge \neg Lk$ Kim the Cretan is not always a liar—(s)he sometimes tells the truth.

Kim the Cretan is obviously a made-up name, used to represent some arbitrary Cretan who is not always a liar. What's surprising is that by saying that all Cretans are always liars, Epimenides, who is from Crete, is thereby really asserting: *some Cretan sometimes tells the truth* (hypothetically, Kim)!

Epimenides wrote only that Cretans are always liars. What if he had made the stronger claim, like that of Eubulides, "I am lying"? If Epimenides had written "I am lying," then a full-fledged liar paradox would result. For, if he's telling the truth when he wrote "I am lying," he must be lying. And, if he was lying when he wrote "I am lying," he must be telling the truth.

An ancient Greek who found the liar paradox most disturbing was Philetas of Cos.

Figure 13.2 Photograph of a bust of Philetas of Cos (ca. 340–285 BCE). Photo credit: White Images / Scala / Art Resource, New York.

Philetas of Cos (also spelled Philitas) was so obsessed with the liar paradox that his intense preoccupation with it supposedly led to his death. Philetas is said to have written the following poem:[15]

> Philetas of Cos am I,
> 'Twas the Liar who made me die,
> And the bad nights caused thereby.

Philetas became so thin pondering the paradox that he tied weights to his shoes so he wouldn't be blown away by the wind. According to some ancient accounts, he actually did die by wasting away (not that we should necessarily believe this).

A way of refining this first-person liar paradox is to write: "This sentence is false."

We can also name the sentence, then write that the sentence with that name is false.

We'll name the sentence "$\mathcal{F}$":

$$\mathcal{F} \qquad\qquad \mathcal{F} \text{ is False.}$$

Is $\mathcal{F}$ true or false?

If $\mathcal{F}$ is true, then since $\mathcal{F}$ names, or is short for, "$\mathcal{F}$ is false," $\mathcal{F}$ must be false.

But if the sentence $\mathcal{F}$ is false, then the sentence "$\mathcal{F}$ is false" must be true.

So, if true, false; and if false, true. Paradox.

Figure 13.3 Etching of a bust of Zeno of Elea (ca. 495–430 BCE). (Etching by Jan de Bisschop from *Paradigmata Graphices Variorum Artificum* / Mr. Nostalgic / Wikimedia Commons.)

Let's cover some other clever paradoxes.

"Achilles and the tortoise" is one of many paradoxes by Zeno of Elea, who designed them to support the philosophy of Parmenides. Parmenides argued that the senses could not be trusted, change is an illusion, and motion is impossible.

Achilles and the Tortoise

Let the fleet-footed hero of the Trojan war Achilles and a tortoise race a given distance, 100 meters for specificity. We'll suppose that Achilles can finish this race in 10 seconds, running 100 times as fast as the slow-moving tortoise. Achilles, in an act of generosity, we will suppose, gives the sluggish tortoise a 10-meter head start to make the race more sporting. Who will win?

If you answered, "Achilles," consider this: when Achilles has caught up with where the tortoise began the race, 10 meters from the start, the tortoise has lumbered 1/10th meter ahead of Achilles (1/100th of Achilles' distance). When

Achilles catches up to that distance, the slow-moving tortoise has again trudged 1/100th of that distance ahead: namely, $1/10 \cdot 1/100 = 1/1000$ meters. And again, when swift Achilles catches up to that distance, the slothful tortoise has plodded his way another hundredth of it. Whenever Achilles catches up to where the tortoise was at any given point, the tortoise has surged 1/100th of that distance ahead. Thus, Achilles never catches the tortoise.

Figure 13.4 is a diagram showing Achilles always lagging behind the trudging tortoise:

Figure 13.4   Illustration of Achilles chasing a tortoise. Image credit: Ernesto Mora.

We depart from this paradox, believing that the reader can readily demystify it.

After you've exposed the faulty reasoning that leads to the paradox, recall that according to Parmenides, all motion is impossible. Thus, strictly speaking, neither the sluggish tortoise nor speedy Achilles would even leave the starting gate. In more modern times, we've defied Parmenides by putting tortoises in orbit around the moon. They were the first earthlings to make the trip and return.

The Paradox of the Barber

Exercise 21: In a town lives a barber who shaves all men who do not shave themselves. The question is: Who shaves the barber? Puzzle it over before turning to a solution at the end of the book. You may be surprised.

The Heterological Paradox

Here's another paradox, attributed to Kurt Grelling, a 20th century German logician. Define "heterological" as an adjective that does not describe itself. "Short" does describe itself since the adjective "short" is itself short. "Long" does not describe itself since the word "long" is not long. The question is whether "heterological" is itself heterological or not. If it describes itself then it doesn't, and if it doesn't then it does—a genuine paradox. This is a semantic paradox, which differs from the other types we consider.

The Voting Paradox

Figure 13.5 Photograph of Martin Gardner (1914–2010). Photo credit: Author: Konrad Jacobs / Source: Archives of Mathematisches Forschungsinstitut Oberwolfach.

The voting paradox can be found in the fascinating book *Aha! Gotcha*[16] by the prolific science writer Martin Gardner, whose "Mathematical Games" column ran in *Scientific American* for more than 25 years. Computer scientist Donald Knuth wrote that "more people have probably learned more good mathematical ideas from Martin Gardner than from any other person in the history of the world, in spite of (or perhaps because of) the fact that he claimed not to be a mathematician himself."[17] John Horton Conway, one of the most revered living mathematicians until his death in 2020, said Gardner "was the most learned man I ever met."[18] Many more accolades for Martin Gardner can be found in the *Notices of the American Mathematical Society*.

Here's how Gardner presents the voting paradox:[16]

> Suppose that three persons—Abel, Burns, and Clark—are running for president. A poll shows that 2/3 of the voters prefer A to B, and 2/3 prefer B to C. Will most voters prefer A to C? Not necessarily! If voters rank the candidates..., a startling paradox arises.

> This paradox, which goes back to the 18th century [to Nicolas de Condorcet], is a famous example of a non-transitive relation that can arise when people make pairwise choices. The concept of transitivity applies to such relations as "taller than," "greater than," "less than," "equals," "earlier than," and "heavier than." In general, when a relation R that holds for xRy and yRz also holds for xRz, the relation is said to be transitive.

> The voting paradox boggles the mind because we expect that the relation prefers always to be transitive. If someone prefers A to B, and B to C, we naturally expect him or her to prefer A to C. The paradox shows that this is not always the case.

Consult this diagram:

> Voter 1: A B C
> Voter 2: B C A
> Voter 3: C A B

Reading from left to right, two voters prefer A to B (V_1 and V_3), and two voters prefer B to C (V_1 and V_2). Yet, notice that two voters, V_2 and V_3, prefer C to A! Is this result because the voting is intransitive?

No! The voting is transitive, yet we can see that the paradox still remains. Why? Let p_1/q_1 represent the fraction of voters preferring A to B, and p_2/q_2 the fraction of voters preferring B to C. We conclude that $[(p_1/q_1 + p_2/q_2) - 1]$ represents the fraction of voters who must prefer A to C, assuming transitivity. Applying this formula to the example, $[(2/3 + 2/3) - 1] = 1/3$. So, given transitivity, only 1/3 of the voters must prefer A to C.

By way of further example, then, if 2/3 of the voters prefer A to B, and 3/4 of the voters prefer B to C, then at least 5/12—(2/3 + 3/4) – 1—must prefer A to C, *assuming transitivity holds*. If transitivity does not hold, anything goes. Now, try this easy exercise.

Exercise 22: 5/12 of voters prefer A to B, and 7/12 prefer B to C. How many must prefer A to C (assuming transitivity)?

Chapter 14
Fallacies

We previously defined a paradox as a statement that, despite apparently sound reasoning from acceptable premises, leads to a conclusion that seems impossible. A fallacy is neither logically valid nor sound. It includes errors in reasoning or a failure in basic logical deduction. Nevertheless, fallacies can be enlightening and can help us better understand complicated problems in logic and mathematics.

Here's an amusing fallacy that can be found on the Internet. First, we pose the question: What's the infinite sum of all the natural numbers? That is, what is the infinite sum of N, where

$$N = 1 + 2 + 3 + 4 + 5 + 6 + 7 + \dots$$

Warning: In the calculations below, one or more serious errors have been committed. See whether you can spot it (or them).

First, let $S = 1 - 2 + 3 - 4 + 5 - 6 + 7 - 8\dots$.
Then, add S to itself to get $2S$ by doing this:

$$
\begin{aligned}
S &= +1 - 2 + 3 - 4 + 5 - 6 + 7 - 8\dots \\
+ S &= \quad\;\; +1 - 2 + 3 - 4 + 5 - 6 + 7\dots \text{(pushing numbers to the right)} \\
\hline
2S &= +1 - 1 + 1 - 1 + 1 - 1 + 1 - 1
\end{aligned}
$$

To sum these alternating plus-and-minus ones, notice that if the sequence were to end with a plus one, $2S = 1$. But if the sequence were to end with a minus one, $2S = 0$. We wish the sequence to continue forever, so we average 1 and 0. Therefore, $2S = 1/2$.

$$2S = 1/2$$
$$\therefore S = 1/4$$

$$N - S = \begin{array}{l} 1 + 2 + 3 + 4 + 5 + 6 + 7 + \dots \\ - \left[1 - 2 + 3 - 4 + 5 - 6 + 7 - \dots \right] \end{array}$$
$$\overline{\; 0 + 4 + 0 + 8 + 0 + 12 + 0 \dots}$$
$$= \quad 4 \left[+1 \quad +2 \quad +3 \; + \dots \right]$$

i.e., $N - S = 4N$.
And from above: $S = 1/4$.
So, $N - 1/4 = 4N$.
$-1/4 = 3N$
$\therefore N = -1/12$

Now, this particular result is fallacious. But some other results that interpret the number sequence differently are not. For example, in string theory, Riemann's zeta function yields this answer: $\zeta(-1) = -1/12$, which is a counterpart of the infinite summation of the natural numbers.

Figure 14.1   Photograph of Srinivasa Ramanujan (1887–1920). Photo credit: Oberwolfach Photo Collection (author unknown) / Wikimedia Commons.

Srinivasa Ramanujan was an unschooled mathematician who arrived at surprisingly sophisticated mathematical results. G. H. Hardy, one of the most prominent English mathematicians of his era, was surprised one day when he received in the mail mathematical notes from Ramanujan. They were almost incoherent. Hardy worked on them for a short time and then announced that they were the product of a mathematical genius. Hardy was later to invite Ramanujan to England, where the two men would collaborate. They published many results together. But always, the senior master Hardy recognized the mathematical superiority of Ramanujan. One day, while Ramanujan lay dying in a hospital bed, Hardy visited him. Announcing that the license number on his taxicab was a dull one, 1729, Ramanujan demurred. Quite the contrary, Ramanujan said. 1729 is a most interesting number, being the smallest number that's the sum of two cubes in two different ways.

Exercise 23: Find the two sets of cubes that each sum to 1729.

One of Ramanujan's results is completely counterintuitive. It involves a technique he invented, called the "Ramanujan summation," which assigns a value to the sum of a type of infinite series.

We wish to obtain a Ramanujan summation to the effect that $1 + 2 + 3 + 4 + 5 \ldots = -1/12$.

To sum an infinite series, we will use Cesàro summations, which assign values to a non-convergent series. That is, the sums we arrive at are not permitted ordinarily, but under certain conditions we can get a Cesàro sum.

First, we obtain two Cesàro sums:

1. $1 - 1 + 1 - 1 + 1 - 1 \ldots = 1/2$.
2. $1 - 2 + 3 - 4 + 5 - 6 \ldots = 1/4$.

The way we arrive at the first sum is to set the infinite series of $+1$ and -1 to A; i.e., $A = 1 - 1 + 1 - 1 + 1 - 1 + 1 - 1$

Then, we subtract A from 1, arriving at $1 - A$ on the left side. We then group the right side, substituting $(1 - 1 + 1 - 1 + 1 - 1 \ldots)$ for A.

The equation now looks like this:

$$1 - A = 1 - (1 - 1 + 1 - 1 + 1 - 1 \ldots)$$

Now, just remove the parentheses on the right-hand side of the equation to get

$$1 - A = 1 - 1 + 1 - 1 + 1 - 1 + 1 - 1 \ldots$$

Since the right-hand side is just A, we substitute A for it and get

$$1 - A = A$$

Now, add A to both sides:

$1 - A + A = A + A$

$-A + A$ cancels out, leaving us with

$1 = 2A$, or
$1/2 = A$

Now, we prove $1 - 2 + 3 - 4 + 5 - 6... = 1/4$.

First, let the above sequence equal B.

$A - B = (1 - 1 + 1 - 1 + 1 - 1...) - (1 - 2 + 3 - 4 + 5 - 6...)$

Removing the right parentheses, we get

$A - B = (1 - 1 + 1 - 1 + 1 - 1...) - 1 + 2 - 3 + 4 - 5 + 6...$

Rearrange the terms on the right to get

$A - B = (1 - 1) + (-1 + 2) + (1 - 3) + (-1 + 4) + (1 - 5) + (-1 + 6)...$

Simplifying the right-hand side:

$A - B = 0 + 1 - 2 + 3 - 4 + 5...$

Substituting B for the right-hand side gives us

$A - B = B$
$A = 2B$

Remembering that $A = 1/2$, we get

$1/2 = 2B$
$1/4 = B$

Now, let $C = 1 + 2 + 3 + 4 + 5 + 6...$

Subtract C from B:

$B - C = (1 - 2 + 3 - 4 + 5 - 6...) - (1 + 2 + 3 + 4 + 5 + 6...)$

Removing the right parentheses:

$B - C = (1 - 2 + 3 - 4 + 5 - 6...) - 1 - 2 - 3 - 4 - 5 - 6...$

Rearranging:

$$B - C = (1 - 1) + (-2 - 2) + (3 - 3) + (-4 - 4) + (5 - 5) + (-6 - 6)\ldots$$
$$B - C = 0 - 4 + 0 - 8 + 0 - 12\ldots$$
$$B - C = -4 - 8 - 12\ldots$$

Now, factor out 4 from the right-hand side:

$$B - C = -4(1 + 2 + 3\ldots)$$

Since $C = 1 + 2 + 3 + 4\ldots$,

$$B - C = -4C$$
$$B = -3C$$

Since $B = 1/4$,

$$1/4 = -3C$$

Thus, $C = -1/12$.

This result is important in the version of string theory called Bosonic string theory.

Ramanujan summation is also important in the Casimir effect in general physics.

To oversimplify, in certain specialized areas of science where physical meanings are attached to numbers, it is surprisingly true that the sum of the infinitely many natural numbers is $-1/12$. But purely mathematically, this is false.

Chapter 15
The Problem with Rabbits

The problem with rabbits is that they multiply fast. Suppose that when two rabbits are born, there's one male and one female. After a month of maturation, the male and female are capable of mating—incest does not matter—and we assume that they always mate successfully, always produce another male–female pair, and never die. After a month of gestation, they give birth and immediately mate again, so that they keep on giving birth month after month. Their offspring of one male and one female must wait a month to mature before they also begin to reproduce. Figure 15.1 is a chart of these rabbits' reproductive pattern.

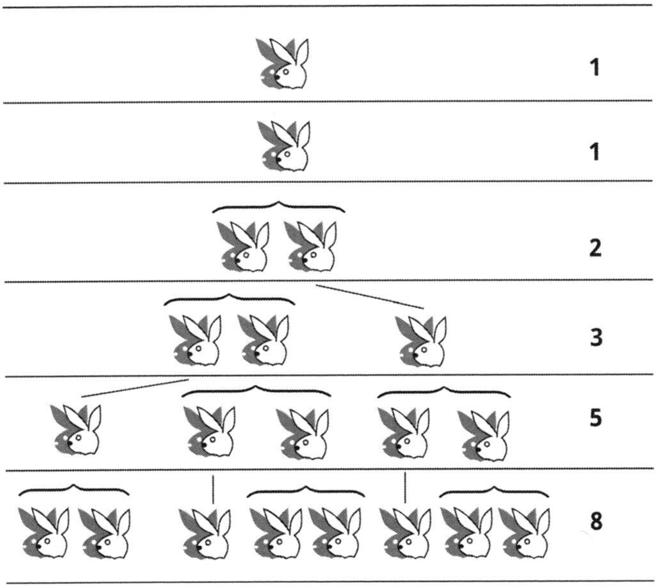

Figure 15.1 Illustration of rabbit reproduction. Image credit: Ernesto Mora.

Exercise 24: How many rabbit pairs appear at the next level? Then the one after that? What is the generalized formula for finding the number of rabbits at any level?

Fibonacci and the Rabbit Challenge

In the year 1202, Fibonacci tackled the rabbit problem presented above. His task was to figure out the number of rabbit pairs after one year. Figure 15.1 shows the breeding pattern for the first six months: 1, 1, 2, 3, 5, 8.

The Fibonacci triangle (Fig. 15.2), representing the reproduction pattern of the rabbits, is just one aspect of Pascal's triangle, shown here:

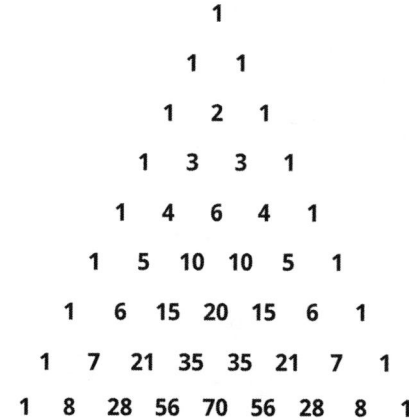

Figure 15.2   Illustration of the Fibonacci triangle. Image credit: Ernesto Mora.

Now, take this triangle and draw diagonals, as indicated in Fig. 15.3. The sum of these diagonal numbers represents the Fibonacci series from 1 to 34.

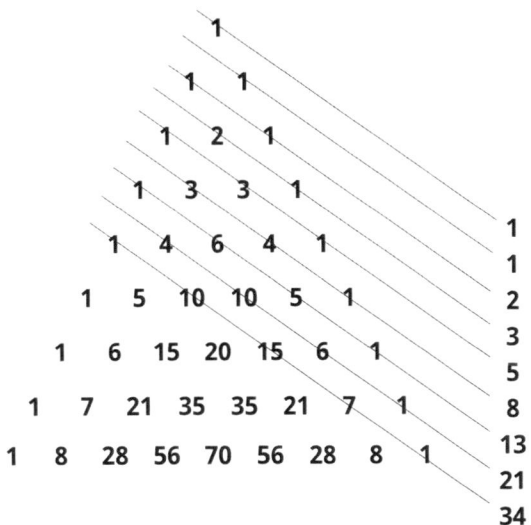

Figure 15.3 Illustration of the Fibonacci triangle with diagonals drawn. Image credit: Ernesto Mora.

Exercise 25: In pseudocode, write a short program in which the Fibonacci numbers are listed from 1 to 100.

The Golden Ratio

Coinciding with the Fibonacci series is the "golden ratio," also called the "golden mean," the "golden section," the "divine proportion," etc. The golden ratio—GR for short—is often symbolized using the Greek letter φ (phi). Look at the squares in the rectangle in Fig. 15.4:

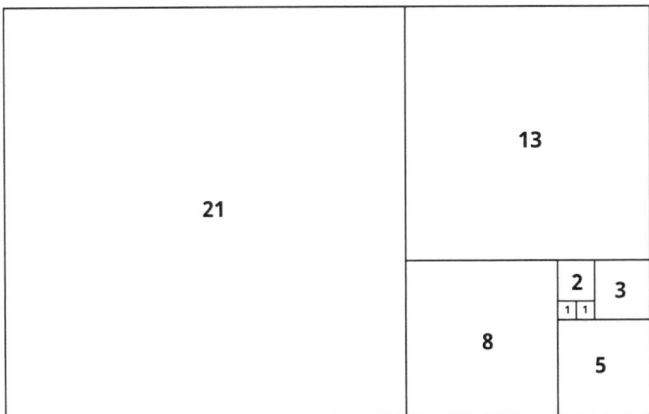

Figure 15.4 Squares illustrating the golden ratio. Image credit: Ernesto Mora.

Going from small to large (and excluding the 1s), the squares are: 2, 3, 5, 8, 13, and 21—the first few numbers of the Fibonacci series. The ratios of two adjacent numbers in this series, from the second to the first, the third to the second, the fourth to the third, the fifth to the fourth, and the sixth to the fifth, are 3/2, 5/3, 8/5, 13/8, 21/13. As these ratios progress, they approach the GR, where φ is the limit.

For example, consider both 13/8 and 21/13. The first ratio, 13/8, is approximately equal to ($\approx$) 1.6250. The second, 21/13, $\approx$ 1.6153. The difference between them is 0.0097, making the two very close. Now, let's solve φ exactly:

> Let a equal the number immediately after b in the sequence.
> Then $\varphi = a/b$, and also $\varphi = (a + b)/a$.
> $a/b = (a + b)/a$; $\varphi = a/a + b/a = 1 + 1/\varphi$.
> So, $\varphi = 1 + 1/\varphi$. Multiplying both sides by φ,
> $\varphi^2 = \varphi(1 + 1/\varphi) = \varphi + 1$. So,
> $\varphi^2 - \varphi - 1 = 0$.
> Now, using the binomial theorem or "completing the square," we arrive at
> $\varphi = (1 \pm \sqrt{5})/2$.
> We want only the positive value. So, $\varphi = (1 + \sqrt{5})/2 \approx 1.6180339$ (a little less than 13/8, and a little more than 21/13).

The golden ratio is found in math (obviously), geometry, nature, architecture, financial markets, art, etc. It's just about anywhere you can think to look.

The GR is shown in Michelangelo's famous painting *The Creation of Adam* (Fig. 15.5). The ratio of its width to its height is similar to the ratio of the width to the height in the rectangle above, measuring 34 by 21, which is the ratio of the ninth number to the eighth number in the Fibonacci series.

Figure 15.5 *The Creation of Adam.* Fresco by Michelangelo Buonarroti / Wikimedia Commons.

Now look at the painting by Salvador Dalí, *The Sacrament of the Last Supper* (Fig. 15.6). The ratio of the width-to-height in Dalí's painting is similar to the ratio in Michelangelo's painting of *The Creation of Adam*: $34/21 \approx \varphi$.

Figure 15.6 *The Sacrament of the Last Supper*. Oil on canvas by Salvador Dalí, 1955 / Bridgeman Images.

There are many, many more examples of the golden ratio in our lives. If you think about this a bit, you'll see it everywhere. An $8'' \times 11''$ piece of paper is close, as are standard computer and TV screens.

It should be noted that Fibonacci did not invent the series that bears his name. Pingala an ancient Indian musical theorist who flourished, according to one source,[30] between the 3rd and 2nd centuries BCE, is said to have stumbled upon both the Fibonacci numbers and binomial coefficients. This discovery is more than 2000 years before Fibonacci. Other scholars from India and Persia, long before Fibonacci, are also claimed to have anticipated his discovery.

Possibly it's going too far with the Fibonacci series to mention one more famous mathematical application, but why should that stop us? Yet another example of the importance of Fibonacci numbers is their use in solving a very famous problem: Hilbert's tenth problem.

Hilbert's Tenth Problem

David Hilbert (Fig. 15.7), addressing the International Congress of Mathematicians convened in Paris in 1900, posed 23 unsolved mathematical problems that he wished to be solved during the 20th century. (Actually, Hilbert only proposed nine problems in Paris; the rest were published later in a transcript.) Many have since been solved or partially solved but others have not. The struggles to solve them, together with the solutions discovered, have shaped modern mathematical thought.

We're going to be looking at Hilbert's tenth problem, and more particularly the role played by Fibonacci numbers in its solution. But first, we need to familiarize ourselves with Diophantine equations.

A Diophantine equation is a polynomial with integer coefficients, usually in several unknowns, where the values of the unknowns are integers and there are fewer equations than unknown variables. For instance, the simplest Diophantine equation has only one unknown, $ax = c$, where the coefficients "a" and "c" are both integers, like this: $2x = 9$. Generally, however, Diophantine equations have at least two unknowns. Take $ax + by = 1$ or $ax^2 + by^2 = cz^2$. Both have solutions among the integers, depending on what integers we take for a and b (and, in the second example, c).

Here's Hilbert's formulation of the tenth problem: "Given a Diophantine equation with any number of unknown quantities and with any number of rational integral numerical coefficients: to devise a process according to which it can be determined by a finite number of operations whether the equation is solvable in rational integers."[19]

Figure 15.7 Photograph of David Hilbert (1862–1943). Photo credit: SUB Göttingen, Sammlung Voit: D. Hilbert, no. 11.

In Hilbert's statement, "rational integral numerical coefficients" just means "integer coefficients" and "rational integers" just means "integers." So, a more modern formulation of Hilbert's tenth (H10, for short) is:

> Provide a procedure which, applied to any given Diophantine equation with integer coefficients, can decide whether the equation has a solution with all unknowns taking integer values.

For example, the Diophantine equation $x^2 + y^2 = z^2$ is the Pythagorean theorem, which has infinitely many primitive triples—3, 4, 5 (and all multiples); 5, 12, 13; 8, 15, 17; etc.

In Hilbert's days, there existed only methods for deciding that a given Diophantine *is* solvable in the integers, not whether one may be *unsolvable*. Hilbert wished to find such a (positive) method. But what if there were no method that would decide all Diophantine equations? We'd be looking forever. That would make the problem "undecidable." What Hilbert lacked when he proposed this problem is a *method* for determining undecidability. As we'll explore in detail in our last chapter, it was only in 1931 that logician Kurt Gödel clearly formulated undecidability.

Figure 15.8 Photograph of Julia Robinson (1919–1985). Photo credit: Author: George M. Bergman / Source: Archives of the Mathematisches Forschungsinstitut Oberwolfach.

Julia Robinson (Fig. 15.8) probably first became enchanted with H10 around 1948, when she received her PhD in mathematics. By 1950, she began an extensive assault on H10. Around the same time, Martin Davis became captivated by the problem before finishing his own PhD in 1950. In 1957, Davis teamed up with Hilary Putnam and together they worked assiduously on H10 day and night. This collaboration resulted in a preprint of an article that they sent to Julia Robinson, who completely revamped their paper and presented novel ideas in a 1961 joint publication. She introduced a startling hypothesis for resolving H10, which Martin Davis dubbed "JR," the initials of her name. (Martin Davis writes that some mathematicians found the JR hypothesis implausible, especially the noteworthy Georg Kriesel.)

For about 20 years the problem languished. Year after year, as Julia Robinson blew out the candles on her birthday cake, she would wish that someone would solve H10 before she died. Then one day, a young Russian mathematician, Yuri Matiyasevich, who had previously given up on H10, was required, somewhat against his will, to review Robinson's paper.

Enter Fibonacci numbers: Matiyasevich noticed that the exponential function required by the JR hypothesis was exhibited in the growth of Fibonacci numbers. Figure 15.9 is a rough graph of the first few Fibonacci numbers. Letting the first number be one, and the second also one, the third is two, the fourth is three, the fifth is five, ..., and the tenth is 55. One can see by the graph that they increase roughly exponentially.

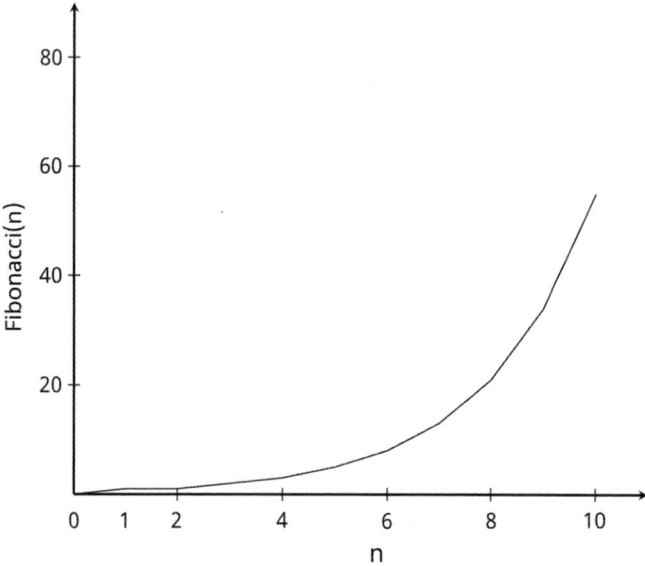

Figure 15.9 Chart illustrating the growth of Fibonacci numbers. Image credit: Ernesto Mora.

Finding an exponential function—the one provided by Fibonacci numbers—that resolved JR, Matiyasevich put to rest Hilbert's tenth problem. Is there a general method for determining whether a given Diophantine equation has a solution in the integers? No! Matiyasevich proved that H10 is "undecidable."

Matiyasevich writes:

> After I read Julia Robinson's paper, I immediately saw that Vorob'ev's theorem could be very useful [in resolving JR]. Julia Robinson did not see the third edition [though she'd seen the first] of Vorob'ev's book until she received a copy from me in 1970. Who can tell what would have happened if Vorob'ev had included his theorem in the first edition of his book? Perhaps Hilbert's tenth problem would have been 'unsolved' [i.e., determined to be undecidable] a decade earlier![20]

Figure 15.10 Photograph of (left to right) Martin Davis, Julia Robinson, and Yuri Matiyasevich in Calgary, Canada, 1982. Photo credit: Yuri Matiyasevich.

After putting a cap on the theorem, Matiyasevich and Robinson collaborated on several papers that improved the basic result. The relationship between Julia Robinson and Yuri Matiyasevich developed into a kind of love story—that of mother and son.

Figure 15.11 Photograph of Hilary Putnam (1926–2016). Photo credit: Hilary Putnam / Wikimedia Commons.

So, in 1970, thanks to Yuri Matiyasevich, Julia Robinson, Martin Davis, and Hilary Putnam (Fig. 15.11), a surprising application of Fibonacci numbers was used to settle this outstanding problem posed by Hilbert in 1900. The resulting theorem is appropriately named MRDP (though the letters are sometimes shuffled).

Chapter 16
Miscellaneous Problems

The remaining chapters in this book are, on the whole, more technical and demanding. Before we go any further, it might be worthwhile (and, I hope, fun) to fool around with some exercises that you should be able to figure out from the previous chapters.

Some of these exercises will be simple. Others may be trickier. Most have solutions in the back of the book.

Travel Problems

Exercise 26: A man drives his car 30 miles per hour for 60 miles in going from Point A to Point B. How fast must he travel on his return journey to average 60 miles per hour for the total distance?

Exercise 27: This one is an oldy but goody. Two trains 100 miles apart are chugging toward each other on the same track. Train A is traveling toward Train B at 20 miles per hour, and Train B is traveling toward train A at 20 miles per hour. A fly flies back and forth between the trains at 40 miles per hour, going from A to B, then B to A, and so forth until the fly gets crushed when the trains collide. How many miles does the fly fly?

American mathematician John von Neumann was once asked to solve a version of this fly–train problem. He gave the answer without a moment's hesitation. "Oh, so you know the trick?" asked the disappointed questioner. Von Neumann replied, "What trick? I just summed the infinite series."

Series

Exercise 28: Here's a problem with a surprising solution. What is the next number in the following sequence:

1, 2, 4, 8, 16, __?

Are you sure? Before answering this question, I should mention that any finite list of numbers that appears to demonstrate a pattern can *in theory* be followed by any number(s). For example, here's an unserious answer to the above question, but a possible answer, nonetheless. The numbers 1, 2, 4, 8, 16, could be followed by 1, 2, 4, 8, 16, 1, 2, 4, 8, 16, etc.

The point is that theoretically one can continue any finite sequence in any conceivable way. Thus, the question raised should be: "What is the likeliest next number after 1, 2, 4, 8, 16?" But even this is open to doubt. There are arguments to be made for multiple "likeliest" next numbers.

Exercise 29: See whether you can extend this interesting sequence: 1, 2, 2, 1, 1, 2, 1, 2, 2, 1, 2, ...

"Two Bottles" Problems

Here is another Diophantine problem, which is of a type we are all familiar with. It's one of the "two bottles" problems. The problem is to find a way to arrive at a bottle containing exactly 6 pints of water if you have two unmarked bottles available—one that can hold 7 pints of water and the other 11. With some cleverness, you can fill and pour water to and from the two bottles until one of them contains the desired 6 pints of water.

The reason this problem is Diophantine is that it can be represented by the equation $7x - 11y = 6$, with two unknowns, though it's only a single equation. This Diophantine equation is relatively easy because it is linear, meaning that there are no exponents above one.

Exercise 30: Solve this "two bottles" problem by juggling the two bottles or by solving the equation and then applying that solution to the problem itself.

Proof of Irrationality

Since we're now considering numerical issues, one problem concerns rational numbers, which can be expressed as fractions. We already looked at the spiral of Theodorus, but we couldn't discern from it how he distinguished rational from irrational numbers. Now, let's try to take things a step further.

The problem is to prove that $\sqrt{2}$ is irrational—that is, not expressible as a fraction, t/b, such that t and b have no factors in common, which is to say that t and b are "co-prime" or "relatively prime" and b is greater than zero.

There are several proofs that there's no fraction equaling $\sqrt{2}$. Here's an indirect proof that uses the fundamental theorem of arithmetic (FTA), which mathematicians often prefer not to use. Fortunately, there are other proofs that do not use the FTA, but for now, we use the FTA, which establishes that every integer greater than 1 can be decomposed—or factored—into primes in a unique way. For example, take any prime, such as 29. Since 29 itself is prime, it's decomposable into primes in a unique way, namely 29. Take a non-prime, such as 27. 27 can be factored into the single number 3, raised to the third power, namely 3^3. For good measure, consider the number 52: $52 = 2^2 \cdot 13$.

Begin the proof by assuming the contrary, that $\sqrt{2}$ is rational, and then prove a contradiction from that assumption, thereby proving that $\sqrt{2}$ is not rational; it must be irrational. Here's a proof of this that uses the FTA:

Prove that $\sqrt{2}$ is irrational:

Assume the opposite that $\sqrt{2}$ is rational. That is:

1. $\sqrt{2} = a/b$, $b \neq 0$ and only 1 divides a and b (i.e., they are co-prime).
2. $2 = a^2/b^2$, from (1), squaring both sides of the equation.
3. $2b^2 = a^2$, multiplying both sides by b^2.
4. There's one single 2 on the left of equation (3), and an even number of unique primes in both a^2 and b^2. Thus, $2b^2$ cannot equal a^2.
5. $\therefore$ $\sqrt{2}$ cannot be rational but must be irrational.

We should explain (4) above. By the FTA, assume that $b = p_1 \cdot p_2 \cdot p_3 \cdot \ldots \cdot p_n$, where all p's are primes. Similarly, suppose that $a = q_1 \cdot q_2 \cdot q_3 \cdot \ldots \cdot q_m$, where all q's are also primes. Thus, $a^2 = (q_1 \cdot q_2 \cdot q_3 \cdot \ldots \cdot q_m) \cdot (q_1 \cdot q_2 \cdot q_3 \cdot \ldots \cdot q_m)$, whereas $2b^2 = 2(p_1 \cdot p_2 \cdot p_3 \cdot \ldots \cdot p_n) \cdot (p_1 \cdot p_2 \cdot p_3 \cdot \ldots \cdot p_n)$. There are pairs of primes in b^2, and similarly there are pairs of primes in a^2. Thus, equation (3) is impossible, since the single 2 on the left makes that side have an odd number of 2s, whereas a^2 on the right must have an even number of 2s (or none at all). This gives us the contradiction in (5). Hence, $\sqrt{2}$ must be irrational.

Here's another indirect proof that shows $\sqrt{2}$ to be irrational without using FTA:

Prove that $\sqrt{2}$ is irrational:

Assume the opposite, that $\sqrt{2}$ is rational. That is:

1. $\sqrt{2} = a/b$, $b \neq 0$ and only 1 divides a and b (i.e., they are co-prime).
2. $2 = a^2/b^2$, from (1), squaring both sides of the equation.

3. $2b^2 = a^2$, multiplying both sides by b^2.
4. $\therefore$ 2 divides a^2, written as "$2|a^2$."
5. Since 2 times something (in this case, b^2) $= a^2$, then
6. $2|a$ also. Let something that divides a be k. So,
7. $2k = a$ from (6). Thus, $(2k)^2 = a^2 = 4k^2$.
8. $2b^2 = 4k^2$ from (3) and (7). Then,
9. $b^2 = 2k^2$, dividing both sides of (8) by 2.
10. But then, $2|b^2$, and thus $2|b$.
11. Let $b = 2q$.
12. Then $a/b = 2k/2q$, from (7) and (11). This is:
13. A contradiction! Since by (1) both a and b were assumed to be in lowest terms. Thus, our assumption was wrong, and
14. $\therefore$ $\sqrt{2}$ is irrational. And we're done.

Here's a very similar proof to the one above, though this one is easier:

1. Suppose that $\sqrt{2} = a/b$, as in (1) in the proof above.
2. Since both a and b cannot be divisible by any number, this is also true of a^2/b^2. In particular,
3. $2 \neq a^2/b^2$, contradicting (3) in the proof above, which proves that
4. $\sqrt{2}$ is irrational.

Now, we'll try to prove that (a) for any $n > 1$, n is a perfect square if $\sqrt{n}$ is rational and (b) if $\sqrt{n}$ is rational, then n is a perfect square. (Together, they prove that n is a perfect square if and only if $\sqrt{n}$ is rational.)

Proof of (a)—we'll leave it to you to fill in any missing steps:

We'll prove the contrapositive: if $\sqrt{n}$ is irrational, then n is not a perfect square:
Assume that $\sqrt{n}$ is irrational and that n is a perfect square. Then $n = m^2$ for some m.
Then $\sqrt{n} = \sqrt{m^2} = m$, and we have. . .
A contradiction!

Proof of (b)—again, fill in any missing steps on your own:
We'll start with the contrapositive again: if n is not a perfect square, then $\sqrt{n}$ is irrational.
$n \neq m^2$ and assume that $\sqrt{n}$ is rational; i.e., $\sqrt{n} = a/b$, where a and b are co-prime.
$n = a^2/b^2$.
$nb^2 = a^2$. Since a and b are co-prime, b must $= 1$. Thus, $n = a^2$.

For fun, you could now prove (along similar lines) that $\sqrt{6}$ is irrational.

Speaking of irrational numbers, there's a theorem to the effect that there are irrational numbers, a and b, such that a^b is rational. Here's a non-constructive proof of that claim. Let x stand for $\sqrt{2}^{\sqrt{2}}$. If we suppose that x is rational, then we're done. So, let x be irrational. Then raise $\sqrt{2}^{\sqrt{2}}$ to the power of $\sqrt{2}$. Then $(\sqrt{2}^{\sqrt{2}})^{\sqrt{2}} = \sqrt{2}^{(\sqrt{2}\cdot\sqrt{2})} = \sqrt{2}^{2} = 2$. So, in both cases, a^b is rational, and we're done. Short and maybe somewhat sweet.

Chapter 17
Unsolved Problems

The Twin Prime Conjecture

We know that only one even number is prime—namely, 2. Thus, all prime numbers greater than 2 are odd. Some pairs of primes, called twin primes, are separated only by a single even number. Here are some examples: 3 and 5, 5 and 7, 11 and 13, 17 and 19, 29 and 31. The question is whether there are infinitely many of these "twins." To state the conjecture positively: there are infinitely many twin primes greater than 2.

Sometimes the twin prime conjecture is attributed to Euclid, though it has never been located in Euclid's writings. The earliest known statement of it comes from 19th century French mathematician Alphonse de Polignac, in a work entitled *Six Arithmetical Propositions Deduced from the Sieve of Eratosthenes* and published in 1849.

For more than 150 years, number theorists have worked diligently on the twin prime conjecture without finding a solution. However, in 2013, a sensational breakthrough was arrived at by an unheralded calculus teacher at the University of New Hampshire. Working in complete isolation, Yitang ("Tom") Zhang (Fig. 17.1) reduced the problem of finding infinitely many primes separated by 2 by proving that there are infinitely many primes separated by 70 million or less. To express this with some attempt at more mathematical precision, the twin prime conjecture is that there are infinitely many primes such that the larger one, a, minus the smaller one, b, equals 2: $a - b = 2$. Yitang Zhang proved that there are infinitely many primes such that $a - b \leq 70,000,000$.

Figure 17.1   Photograph of Yitang Zhang. Photo credit: Voice of America.

Yitang Zhang's result may not seem to be such a big deal since 70 million is such a large number, but his accomplishment was a tremendous breakthrough. Since his result was published, others have applied some of his concepts in the search for smaller bounds. For instance, James Maynard gave a different proof of Zhang's theorem, thereby reducing the gap from 70,000,000 to 600. The Polymath Project, a group of mathematicians who collaborate online, subsequently reduced the gap to 246. Given various mathematical conjectures, the gap can be further reduced. But, without assuming anything, the gap remains, as of this writing, at 246. The research continues. . .

Goldbach's Conjecture

We consider Goldbach's conjecture (GC) to be: "Every even number greater than 4 is the sum of two odd primes." There's a more familiar version of GC, but we'll leave that to the curious reader to explore.

Let's look at the first three examples:

$$6 = 3 + 3$$
$$8 = 5 + 3$$
$$10 = 7 + 3$$

If it were to turn out that GC is independent (not provable or disprovable) in some sufficiently strong mathematical theory such as Peano arithmetic, then it would be true. For if "no" were the case, then we would have a counterexample to GC by just exhibiting the even number. German mathematician Christian Goldbach proposed this problem in 1742 in a letter sent to the brilliant polymath Leonhard Euler. It's one of the most famous unsolved problems in mathematics. Computer scientist Donald Knuth has stated that maybe GC is undecidable.

Let's look at one version of the weak Goldbach conjecture (WGC). It states that every odd number greater than 7 is the sum of three odd primes. GC proves the WGC, as we can just add 3 to the even numbers greater than 4 and get odd numbers greater than 7. For instance: $6 + 3 = 9$, $8 + 3 = 11$, etc. The WGC was proved by Harald Helfgott in 2013. It should be mentioned that just plain GC has been found to be true by computer search for numbers up to 10^{18}.

Legendre's Conjecture

An unsolved problem that looks as though it must be true is Legendre's conjecture. It states that for any natural number (n) that is greater than 1 ($n > 1$), there must be a prime between n^2 and $(n + 1)^2$. Let's look at some small numbers, starting with 2. Between 2^2 and 3^2 are both 5 and 7. Now, let n be 3. Between 3^2 and 4^2 are 11 and 13. For $n = 4$, $4^2 = 16$ and $5^2 = 25$, making 17, 19, and 23 work. At this juncture, it appears that the number of primes may actually increase between n^2 and $(n + 1)^2$. However, Legendre's conjecture has so far evaded proof.

The Steiner–Lehmus Theorem

The Steiner–Lehmus theorem is an intriguing and seemingly simple problem in plane geometry. It is named after Jakob Steiner (1796–1868) and C. L. Lehmus (1780–1863). Lehmus originated the problem, which was passed on to Steiner, who was the first person to solve it using purely elementary geometric means. His solution, however, is equivalent to a *reductio ad absurdum* proof, or an indirect proof.

The problem is shown in Fig. 17.2.

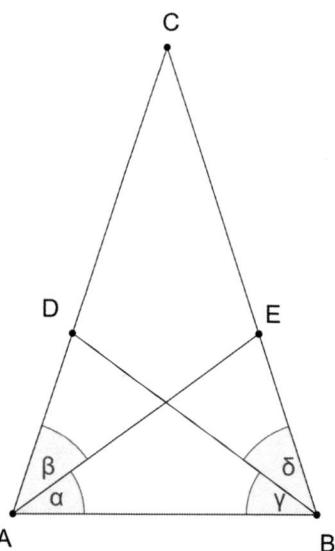

Figure 17.2 Diagram of the Steiner–Lehmus problem. Image credit: Kmhkmh / Wikimedia Commons.

Given triangle ABC, with the length of the angle bisector AE equal to the length of the angle bisector BD (i.e., α = β and δ = γ), prove that triangle ABC is isosceles by a direct proof using purely geometric means. Both Steiner and Lehmus were able to furnish elementary geometric proofs, but not direct ones. Indeed, it has been claimed that no such direct proof exists.

To tantalize you, here's a supposedly direct geometric proof. In Fig. 17.3 (note that it differs from the one above), we've drawn both GE and DF parallel and equal to AB. Argue that parallelogram BAEG is congruent to BAFD. Initially, this seems to work, making the base angles CAB and CBA equal, rendering triangle ABC isosceles. What do you think? Is this okay?

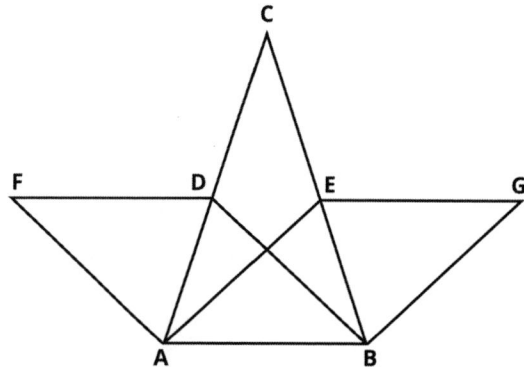

Figure 17.3 Diagram of a supposedly direct geometric proof of the Steiner–Lehmus problem. Image credit: Ernesto Mora.

Nope! One way to see this is that we don't know that line segment AD equals BE. To convince yourself that this matters, construct two *unequal* parallelograms BAEG and BAFD, meeting all the criteria.

Professor John Horton Conway has written, "if there were a proof of the Steiner–Lehmus theorem of the ... equality-chasing type, it would continue to work when we varied the triangle smoothly ... and so would prove this triangle equilateral, which it isn't. So, there's no such proof!"[21] Unfortunately, I haven't an inkling of what Conway means by "a proof of the equality-chasing type." However, you're free to see whether you can find an elementary direct geometric proof of this innocuous-looking theorem—i.e., a proof that doesn't rely on any indirect arguments.

The Midpoint Theorem

A brief departure from unsolved problems seems warranted while we're on the topic of plane geometry. Let's play with an interesting and not too difficult theorem, called the midpoint theorem. The midpoint theorem states that the line segment in a triangle joining the midpoint of two sides is parallel to its third side and is also half of the length of the third side.

Exercise 31: Prove the midpoint theorem—i.e., in any triangle, the line joining the midpoint of two sides is parallel to the third side and equal to one half of it.

Exercise 32: After proving the midpoint theorem in the previous exercise, prove the "converse" of the midpoint theorem. That is, if a line segment passes through the midpoint of one side of a triangle and is parallel to another side, then it bisects the third side. (This is not the literal converse of the theorem, which would be: if a line intersecting two sides of a triangle is parallel to the third side and equal to one half of it, then the intersection line is the midpoint of the first two sides.)

Exercise 33: Slightly modify the proof of the theorem called (in the previous exercise) the "converse" of the midpoint theorem to obtain the proof of the literal converse of the theorem.

The Collatz Conjecture

Returning to unsolved problems, we have the Collatz conjecture or the $3n + 1$ conjecture.

Pick a number—make it 5, just as an example. 5 is odd, so multiply it by 3. That's 15. Add 1. That's 16. Since 16 is even, divide by 2. That's 8, so divide by 2 again, which is 4. 4 divided by 2 is 2, and 2 divided by 2 equals 1.

Now, start with an even number (that's not a power of 2)—make it 12. 12 is even, so divide by 2. That's 6. Divided by 2 equals 3. 3 is odd, so multiply by 3 and add 1. That's 10. Divided by 2 equals 5. And we saw above that from 5, we arrive at 1 after performing successive operations.

More generally, the $3n + 1$ conjecture is this: pick any positive integer and call it n. If n is odd, multiply it by 3 and add 1. If n is even, divide n by 2. The conjecture is that performing successive operations on these resultant numbers will always lead to 1.

Here is the cycle of numbers beginning with 15, and performing the indicated operations: 15, 46, 23, 70, 35, 106, 53, 160, 80, 40, 20, 10, 5, 16, 8, 4, 2, 1.

Try out some other numbers. You'll always arrive at 1. At least, this is the conjecture. It has been tested for huge numbers, so you aren't likely to find a counterexample. But so far it hasn't been proven true for all numbers. The great mathematician Paul Erdős said around 1990 that the problem was too difficult for solution in the near future.

More Diophantine Equations

The following decidable problem is solvable, though solving it can be tedious if you just try out possible solutions. The problem is this: find two positive whole numbers such that the first one cubed is two more than the second one squared. One way to write an equation for it is $x^3 - y^2 = 2$.

Notice that there are two variables and only one equation, which we have learned in beginning algebra cannot be solved. As we have already seen in the section on Hilbert's tenth problem, this is untrue. Solutions to many such equations, called Diophantine, can be arrived at by using various techniques. For Diophantine equations in general, there is no single technique for solving them.

Without using any technique, we can just plug in some small numbers and see what works. For instance, if we restrict our attention to positive integers, then $3^3 - 5^2 = 2$ works, which may be the only solution for positive integers.

Chapter 18
Sentential Logic

Sentential logic (SL) is the formal logic of complete sentences. Some of the infinitely many sentence letters of sentential logic are

$$P, Q, R, S, P_1, Q_1, R_1, S_1, \ldots P_{325}, Q_{325}, R_{325}, S_{325}, \ldots, P_{1000}, \ldots$$

The letters, with and without subscripts, are called atomic sentences of SL. Some of the more familiar connectives are the two-placed (dyadic) connectives. We introduced these to you way back in Chapter 5, but we'll reintroduce them here:

∧	*And*
∨	*Or*
→	*If..., then...*
↔	*If and only if* (sometimes written "iff")

The most familiar one-placed (monadic) connective is:

¬	*Not*

We'll first present a recursive definition of just the "simple sentences" of SL— i.e., those only with → and ¬. The letters listed above are the atomic sentences of SL. We'll use the Greek symbols φ (phi) and ψ (psi) as variables to stand for "any sentence" of SL. Now, we'll define "simple sentences" of SL.

1. All atomic sentences are simple sentences of SL.
2. For any sentence φ of SL, ¬φ is also a simple sentence of SL.
3. For any two sentences φ and ψ of SL, (φ → ψ) is a simple sentence of SL.
4. Nothing else is a simple sentence of SL.

It is unnecessary to write clause 4, the "That's all, folks!" clause, because it's understood. Although sometimes it is included just to make sure it's understood.

Here are just a few of the infinitely many simple sentences that use just P and $\rightarrow$:

$(P \rightarrow P)$, $(P \rightarrow (P \rightarrow P))$, $((P \rightarrow P) \rightarrow P)$, ...

Exercise 34: First note that two of the above simple sentences have three occurrences of P. Write at least four more simple sentences with just P and $\rightarrow$ that use exactly five occurrences of the atomic sentence P. How many of this type are there in total? On your own and just for fun, try to write a recursive definition for *all* sentences of SL—not just the simple sentences. For instance, $(P \wedge Q)$ is a sentence of SL.

I'm sure you know the two truth values, True and False, or just T and F. And you probably know most if not all of the standard truth functions. Let's look at the definitions of a few truth functions based on the above connectives.

The simplest connective is the monadic one, "not." It gives us this straightforward truth table "not-True is False" and "not-False is True." Using the (sentential) letter P, we have:

P	$\neg P$
T	F
F	T

"Not" isn't the only truth function with a single sentential letter; it's just the most familiar one. Actually, there are four truth functions for a single sentence. Below, they are all represented (though the only one in which we're interested is $\neg P$):

P	$\%P$	$*P$	$\neg P$	$\#P$
T	T	T	F	F
F	T	F	T	F

Let's look at the truth table for "or." If either or both of two sentences, P or Q, is T, then the sentence $(P \vee Q)$ is True; otherwise not. This is called the inclusive sense of "or," since it includes the possibility that both P and Q are True together.

P	Q	$P \vee Q$
T	T	T
T	F	T
F	T	T
F	F	F

The dyadic or two-placed connective "and" is unsurprising; if both P and Q are True, then $(P \wedge Q)$ is True. Otherwise, it's False. Here is its truth table:

P	Q	$P \wedge Q$
T	T	T
T	F	F
F	T	F
F	F	F

Of all the familiar truth functions, the only really surprising one belongs to the connective "if..., then..." In logic, the only time a conditional sentence is False is when the "if" part is True and the "then" part is False.

The assignment of truth value for "if..., then..." is not completely intuitive, but consider the following reasoning. Make a bet with someone that if two dice come up six on a roll, then on the next roll each of the dice will display two distinct numbers. Suppose you roll the dice and double sixes do not show up on the first roll. Should we consider this a win or a loss for the person who made the bet? One way to evaluate this possibility is to keep rolling until double sixes do show up, and then see whether on the next throw distinct numbers show up. For our purposes, however, we'll interpret the "if..., then..." as meaning: either double sixes do not show up on the first roll or two distinct numbers do show up on the second.

It is not at all clear that this is what "if..., then..." really means, but this is how we'll decide to interpret a conditional sentence. There are some very good arguments in favor of it (and some against it), and this definition works extremely well in mathematical contexts, so we stick with it. Here's the truth table for the conditional sentence:

P	Q	$P \rightarrow Q$
T	T	T
T	F	F
F	T	T
F	F	T

Because we're going to refer to it later, let's display the truly simple truth table for "if and only if..." ("iff"). It's simple because "P iff Q" is True exactly when they both have the same truth value—i.e., they're both true or both false. If they have opposite truth values, "iff" is False.

P	Q	$P \leftrightarrow Q$
T	T	T
T	F	F
F	T	F
F	F	T

Truth Tables versus Truth Functions

We need to clarify the difference between truth tables and truth functions. For any number of sentence letters, there are several distinct truth tables. It's the entire extent of them—i.e., all possibilities—that constitute the truth functions. For a relatively simple example, let's look at every possible truth function for just two sentence letters, P and Q:

P	Q	1	2	3	4	5	6	7	8	9	10	11	12	13	14	15	16
T	T	T	T	T	T	F	T	T	T	F	F	F	T	F	F	F	F
T	F	T	T	T	F	T	T	F	F	T	T	F	F	T	F	F	F
F	T	T	T	F	T	T	F	T	F	T	F	T	F	F	T	F	F
F	F	T	F	T	T	T	F	F	T	F	T	T	F	F	F	T	F

Notice how many there are! 16 for just two letters. We can identify the truth table column for the connective $\wedge$ above. It is column number 12, indicating that when both P and Q are True, then $P \wedge Q$ is True, otherwise it's False.

For $P \leftrightarrow Q$, look for T, F, F, T when both P and Q have the same truth value. That's shown in column 8.

Exercise 35: For the connective $\rightarrow$ ("if..., then..."), we'll look for the column where P is T and Q is F (and the rest are T). Select the appropriate column. Then, find the column representing the connective $\vee$ ("or").

Isn't it amazing how many possible truth functions there are for just two letters? What if we used three letters, or four? Well, we had better explain how they proliferate.

Here's how we can predict the number of truth functions for some number of letters. Notice that with a single sentential letter, say P, we have only two truth values, T and F, but four truth functions—i.e., 2^2. This was shown above for the letter P. For two letters, there are $(2^2)^2 = 16$ truth functions, as we saw with P and Q, above. For three letters, say P, Q, and R, there are $((2^2)^2)^2 = 256$ truth functions. And so on. Given n letters, there are 2^{2^n} truth functions. Interesting, isn't it? For such a simple language, the exponential growth of truth functions is surely surprising. Considering the vast number of truth functions for sentence letters, it is remarkable that we can write a single sentence in sentential logic that covers all truth tables for any given number of letters.

We'll show this by example. Once you see the technique for just a few sentence letters, you'll see how to extend this technique to any number of sentence letters you have the patience to try out.

What we'll do here is begin with three ordinary connectives and argue that these three connectives are sufficient for representing all of the truth functions for those letters. Then, we'll winnow the three down to two connectives to establish the same thing. But the real surprise is that we can then winnow them down to a single connective alone (actually, two single connectives separately!), which represents all truth functions for any number of letters.

We'll begin with the two dyadic connectives, $\wedge$ ("and") and $\vee$ ("or"), and the one monadic connective, $\neg$ ("not"). We'll show that a sentence having these three truth-functional connectives can represent all truth functions there are for any given number of letters. The more technical term for this kind of sentence is disjunctive normal form (DNF).

It is intuitively clear—or will be shortly—that a sentence in DNF can represent all truth tables for three sentence letters. But you may need to play around a bit to concede that sentences in DNF are truly sufficient. Consider the truth table for the sentence $(P \rightarrow Q) \rightarrow R$:

P	*Q*	*R*	*(P → Q)*	*(P → Q) → R*
T	T	T	T	T
T	T	F	T	F
T	F	T	F	T
T	F	F	F	T
F	T	T	T	T
F	T	F	T	F
F	F	T	T	T
F	F	F	T	F

Here's the sentence in DNF that has exactly the same truth values (in the exact same order) as above:

$$(P \wedge Q \wedge R) \vee (P \wedge Q \wedge \neg R) \vee (P \wedge \neg Q \wedge R) \vee (P \wedge \neg Q \wedge \neg R)$$
$$\vee (\neg P \wedge Q \wedge R) \vee (\neg P \wedge Q \wedge \neg R) \vee (\neg P \wedge \neg Q \wedge R) \vee (\neg P \wedge \neg Q \wedge \neg R)$$

Take the seventh disjunct in the above truth table, for example—when P is F, Q is F, and R is T. You can see that $(P \to Q) \to R$ happens to be True in DNF (on line 7).

Now consider the sixth line in the above truth table. It says that when P is F, Q is T, and R is F, that $(P \to Q) \to R$ is False. Thus, the sentence in DNF which is $(\neg P \land Q \land \neg R)$, is False as well.

If you have the patience to go through all eight cases, you'll see that the sentence in DNF represents the truth value for every line of the sentence $(P \to Q) \to R$. That is, whenever a column of the truth table is T, there's a disjunct in DNF representing a T line of the truth table. And whenever a line of the truth table is F, there is a sentence in DNF that is also F. That this can be done for any truth table establishes the sufficiency of the set of connectives $\{\neg, \land, \lor\}$. We have provided only a single example, but you can try out a few others that interest you. I believe you'll be satisfied that a sentence in DNF can represent all possible truth functions.

It turns out that we can reduce the set of three connectives $\{\neg, \land, \lor\}$ to two connectives. Each pair of connectivities $\{\neg, \land\}$, $\{\neg, \lor\}$, and $\{\neg, \to\}$ is sufficient for determining all truth functions, as the connectives in each of them can define the others. We'll establish just one of them.

We will define away the connective $\land$ using just $\neg$ and $\lor$, showing that $\{\neg, \lor\}$ is sufficient without the $\land$. We do this by defining $(P \land Q)$ as $\neg(\neg P \lor \neg Q)$. Thus, $\land$ is superfluous.

Exercise 36: Similarly, $\{\neg, \land\}$ can define $\lor$. Provide such a definition. Then, on your own, prove that $\{\neg, \to\}$ is a sufficient set of connectives.

Shockingly perhaps, $\{\neg, \leftrightarrow\}$ is not sufficient.

Exercise 37: Prove that $\{\neg, \leftrightarrow\}$ is not sufficient. Warning: it's harder to prove insufficiency than sufficiency. It will require some thought and intuition. But the proof is very neat, slick, nifty, even spiffy—which doesn't mean you'll be able to prove it fast; you probably will need to think a bit. But when you do see it (even if you need to look up the answer), you'll surely agree that it's nifty.

Even more surprisingly, two single connectives—one called the Sheffer stroke but actually proved first by Charles Sanders Peirce, the other called "the dagger," also due to Peirce—are each sufficient all by themselves.

Figure 18.1 Photograph of Charles Sanders Peirce (1839–1914). Photo credit: New York Public Library, The Miriam and Ira D. Wallach Division of Art, Prints and Photographs: Print Collection.

The Sheffer stroke is a vertical line that looks like this: |. "The dagger" looks like this: ↓.

The stroke is often written as NAND, meaning that not both sentences are True. Using ordinary connectives, this would be indicated by $\neg(P \wedge Q)$. The dagger is written as NOR, meaning that neither of two sentences is true—i.e., $\neg(P \vee Q)$.

More simply:

$P \mid Q$ is True just in case ("iff") not both P and Q are True.
$P \downarrow Q$ is True just in case ("iff") both P and Q are False.

Here they are, defined in tables:

P	*Q*	*P | Q*
T	T	F
T	F	T
F	T	T
F	F	T

P	*Q*	*P ↓ Q*
T	T	F
T	F	F
F	T	F
F	F	T

We'll show that the truth table for ¬*P* in terms of the Sheffer stroke and the ¬*P* for the dagger are identical.

P	*¬P*	*P ↓ P*
T	F	T
F	T	T

P	*¬P*	*P | P*
T	F	T
F	T	T

Exercise 38: Show that both the Sheffer stroke and the dagger are each sufficient alone for representing all truth functions.

Rules of Inference

We now present four natural deduction rules of inference for → and ¬ in sentential logic. The four inference rules involve only these two binary connectives.

> **Rule 1.** The Assumption rule (abbreviated As) permits you to write any assumption you wish, as long as you write the line number to the right of the line on which the assumption occurs; this number is called its "assumption number."

> **Rule 2.** *Modus ponens* (abbreviated MP), which was explained earlier, allows you to write Q on a line of a proof if both P and $P → Q$ occur on

two previous lines. You must cite the assumption numbers of both P and $P \to Q$ as assumptions.

Rule 3. Conditional proof (abbreviated CP) allows you to write a sentence of the form $P \to Q$ on a line of a proof if Q alone occurs on a previous line. If P appears on an earlier line with a single assumption number (meaning that P is an assumption), it's permitted to delete that one assumption number from the assumption numbers of the conclusion, $P \to Q$. If more than one assumption accompanies P, you are not permitted to eliminate any assumption numbers from $P \to Q$. To repeat, if Q is on a line and P does not appear previously, you can still write $P \to Q$, but of course, since P does not occur, there is no assumption number to be deleted from the numbers for $P \to Q$. And if P does occur previously with more than one assumption number, then, as indicated, it is not an assumption, and no number can be deleted. Violation of the restriction to delete only the single assumption number when P was entered as an assumption can lead to all sorts of mistakes. In general, these mistakes will allow you to prove anything whatsoever.

Rule 4. *Modus tollens* (abbreviated MT) implies that when any sentence of a conditional form—like this: $(\neg\varphi \to \neg\psi)$—is on a line, you are permitted to enter φ on a later line if ψ or $\neg\neg\psi$ has also appeared on an earlier line (since both sentences ψ and $\neg\neg\psi$ negate $\neg\psi$). Assumption numbers for φ are all assumption numbers of the two previous lines.

A helpful inference rule we've seen already is *reductio ad absurdum* (sometimes abbreviated RAA). Someone may miss having an RAA rule because of its convenience, though technically it isn't needed.

Reductio ad absurdum is a very powerful rule to use when you cannot see another way to proceed with a proof. After you've assumed the negation of a sentence you wish to prove, suppose that two contradictory sentences (φ and $\neg\varphi$) appear on two lines. You are then permitted to derive any sentence ψ that you wish on a later line, using the assumption numbers of both contradictory sentences (φ and $\neg\varphi$). And, if one of the two contradictory sentences happens to be a sentence you wish to prove, then you can delete the single assumption number of that contradictory sentence. As mentioned, the *reductio ad absurdum* rule is strictly superfluous to the other four rules, though it often nicely shortens proofs.

There's another kind of rule, although this one is perhaps obvious and possibly needs no explanation. Take the tautology $P \to (Q \to P)$. Clearly, $Q \to (P \to Q)$ is another example of this same pattern, as well as $(P \to Q) \to ((Q \to R) \to (P \to Q))$. Both are of the same form as $P \to (Q \to P)$.

The second one is obtained from $P \to (Q \to P)$ by the following substitution, where the substituent (the substituted sentence) is below the substituted letter: $P/P \to Q$ and $Q/Q \to R$. Any such substitution for a tautology—and only a tautology—is guaranteed by uniform replacement (UR). But since this rule is so obvious, we won't need to justify it.

What about inference rules for the other connectives $\wedge$ ("and"), $\vee$ ("or"), and $\leftrightarrow$ ("if and only if..." or "iff")? We haven't provided any inference rules for them. There are rules for them, but we don't need them for now.

One form of *modus tollens* (MT) is the following symbolic argument:

1. $P \to Q$
2. $\neg Q$
$\therefore \neg P$

Looking at a truth-table analysis of the above symbolic argument, assume that the conclusion is False. That is, $\neg P$ is False. That means that P must be True. Then, by the first premise, Q must also be True. (If P is True and Q is False, then the first premise is False.) But if Q is True, that contradicts the second premise, according to which $\neg Q$ is True. A contradiction! This shows that (this form of) MT is valid.

Here's an interesting axiomatic basis for proving all valid (tautological) sentences of sentential logic using just the connectives $\to$ and $\neg$. Call the system below AX. Its axioms are

A1. $P \to (Q \to P)$
A2. $((P \to (Q \to R)) \to ((P \to Q) \to (P \to R))$
A3. $(\neg Q \to \neg P) \to (P \to Q)$

Here is a natural deduction proof—i.e., a "derivation"—using just As and CP to get the first axiom above. Note that additional "To Prove" lines simplify the direction of the argument. Assumption numbers come first, followed by line numbers in parentheses.

A1. $P \to (Q \to P)$

To Prove: $P \to (Q \to P)$
1. P 1, As (1)
To Prove: $Q \to P$
2. Q 2, As (2)
3. $Q \to P$ 1, CP (1, 2)
4. $P \to (Q \to P)$ CP (1, 3)

Perhaps the simplest way to show A1 to be a tautology (valid in sentential logic) is by looking at the only way $P \rightarrow (Q \rightarrow P)$ can be False. This happens when the antecedent P is True and the consequent, $Q \rightarrow P$, is False. This occurs only when Q is True and P is False. But, when P is True, it can't be False. Thus $P \rightarrow (Q \rightarrow P)$ must always be True.

Add to those axioms above the single inference rule MP. [Uniform replacement of sentences for sentences of the same form is assumed: e.g., $\neg Q \rightarrow ((R \rightarrow S) \rightarrow \neg Q)$ is also an instance of A1.]

Now, here's a derivation of A2, using both MP and CP. Remember, the assumption numbers come first, followed by line numbers in parentheses.

> To Prove: $((P \rightarrow (Q \rightarrow R)) \rightarrow ((P \rightarrow Q) \rightarrow (P \rightarrow R))$
> 1. $P \rightarrow (Q \rightarrow R)$ 1, As (1)
> To Prove: $(P \rightarrow Q) \rightarrow (P \rightarrow R)$
> 2. $P \rightarrow Q$ 2, As (2)
> To Prove: $P \rightarrow R$
> 3. P 3, As (3)
> To Prove: R
> 4. $Q \rightarrow R$ 1, 3, MP (1, 3)
> 5. Q 2, 3, MP (2, 3)
> 6. R 1, 2, 3, MP (4, 5)
> 7. $P \rightarrow R$ 1, 2, CP (3, 6)
> 8. $(P \rightarrow Q) \rightarrow (P \rightarrow R)$ 1, CP (2, 7)
> 9. $(P \rightarrow (Q \rightarrow R)) \rightarrow ((P \rightarrow Q) \rightarrow (P \rightarrow R))$ CP (1, 8)

Here's a derivation of A3, $(\neg Q \rightarrow \neg P) \rightarrow (P \rightarrow Q)$, using only CP and MT (and As).

> To Prove: $(\neg Q \rightarrow \neg P) \rightarrow (P \rightarrow Q)$
> 1. $\neg Q \rightarrow \neg P$ 1, As (1)
> To Prove: $P \rightarrow Q$
> 2. P 2, As (2)
> 3. Q 1, 2, MT (1, 2)
> 4. $P \rightarrow Q$ 1, CP (2, 3)
> 5. $(\neg Q \rightarrow \neg P) \rightarrow (P \rightarrow Q)$ CP (1, 4)

Here's an axiomatic proof of $P \rightarrow P$ in AX, using only MP (and uniform replacement). The assumption rule (As) is forbidden.

> To Prove: $P \rightarrow P$
> 1. $P \rightarrow ((P \rightarrow P) \rightarrow P)$ A1, UR
> 2. $(P \rightarrow ((P \rightarrow P) \rightarrow P)) \rightarrow ((P \rightarrow (P \rightarrow P)) $ A2, UR $Q/P \rightarrow P, R/P$
> $\rightarrow (P \rightarrow P))$

3. $(P \to (P \to P)) \to (P \to P)$ 1, 2, MP
4. $(P \to (P \to P))$ A1
5. $P \to P$ 3, 4, MP

Note the complexity of the above derivation of $P \to P$ in our axiomatic system AX. By contrast, here's a natural deduction derivation of the same sentence. (In natural deduction, assumptions are permitted, but they are not in AX.)

To Prove: $P \to P$
1. P 1, As (1)
2. $P \to P$ $\varnothing$, CP (1) where $\varnothing$ is the empty set of sentences

Pretty nifty, no?

Here's another proof (not in AX) that uses both MP and CP (and As):

To Prove: $P \to ((P \to Q) \to Q)$
1. P 1, As (1)
To Prove: $(P \to Q) \to Q$
2. $P \to Q$ 2, As (2)
To Prove: Q
3. Q 1, 2, MP (1, 2)
4. $(P \to Q) \to Q$ 1, CP (2, 3) (eliminating Assumption 2)
5. $P \to ((P \to Q) \to Q)$ CP (1, 4) (eliminating Assumption 1)

Here's a wickedly difficult proof of $P \to \neg\neg P$, using just As, CP, and MT. As you can see, the reasoning is so convoluted that even a sequence of "To Prove's" won't help. Relax, no other proofs in (sentential) natural deduction will be this difficult. The purpose here is just to showcase one elaborate proof. Difficult as it is, dare I say that the proof is nifty? Probably not! A more apt description is that it is ingenious. "Annoying" might be another description. Better for you to just go over it once or twice—of course, more times if it interests you—then move on.

To Prove: $P \to \neg\neg P$
1. $\neg\neg\neg P$ 1, As (1)
2. $\neg\neg P$ 2, As (2)
3. $\neg\neg\neg\neg\neg P \to \neg\neg\neg P$ 1, CP (1)
4. $\neg\neg\neg\neg P$ 1, 2, MT (1, 2)
5. $\neg\neg P \to \neg\neg\neg\neg P$ 1, CP (2, 4)
6. $\neg P$ 1, MT (1, 5)
7. $\neg\neg\neg P \to \neg P$ CP (1, 6)
8. P 8, As (8)
9. $\neg\neg P$ 8, MT (7, 8)
10. $P \to \neg\neg P$ CP (8, 9)

Just to show how helpful *reductio ad absurdum* (RAA) can be, look at the following simpler proof with "To Prove's" in it.

To Prove: $P \rightarrow \neg\neg P$

1. P	1, As (1)
2. $\neg P$	2, As (2)

To Prove: Prove a contradiction

3. $\neg\neg P$	1, RAA (delete 2, as: $\neg\neg P$ contradicts $\neg P$)
4. $P \rightarrow \neg\neg P$	CP (from 1 and 3, no assumptions)

Easy, wasn't it? Perhaps so easy that RAA should be explained more carefully. Regarding the above proof, P on line 1 is just the first part of the conditional, called its "antecedent." P was assumed because the basic way to prove a conditional is to assume the antecedent, then try to prove the second part, after the arrow, called the "consequent." But we then assumed the sentence $\neg P$, which contradicts the desired consequent $\neg\neg P$. We assumed this contradictory sentence in order to use RAA. This desire was announced as "To Prove: Prove a contradiction." Well, the contradiction was already there (P and $\neg P$), so we put on the next line (line 3) the consequent we desired. What's important to note carefully is that RAA permitted the deletion of the assumption $\neg P$, since $\neg\neg P$ contradicts $\neg P$. Once we had that, CP finished the job, since it was the antecedent of $P \rightarrow \neg\neg P$, and the assumption of P on line 1 can be dropped (by CP).

Exercise 39: Prove $(\neg\neg P \rightarrow P)$ with RAA. If you want to obtain a more frightening proof than the annoying one above, don't use RAA.

The logical reasoning behind the earlier proof of $(P \rightarrow \neg\neg P)$ without RAA is convoluted and difficult. Please don't be scared off by its complexity: the logic of other proofs will be more transparent. Remember, it's truth we want, although we're now relying on a syntactic proof system consisting of artificial sentences to get us there. But step back and look at the theorem in terms of its truth (which is semantic). It states that $P \rightarrow \neg\neg P$, which is easily seen to be True by a simple truth table.

Truth and Proof

Truth tables provide a semantic analysis of an argument. In a syntactic argument having only the symbols $\neg$ and $\rightarrow$, we use our aforementioned inference rules: As, MP, MT, and CP (and sometimes RAA).

It can be proved that these inference patterns preserve truth, always leading from premises accepted to be True to True conclusions.

We claim that each of our inference rules is valid; taken together the resulting system is sound.

For instance, let's justify the inference pattern MT. Here, again, is one version of it:

1. $P \to Q$
2. $\neg Q$
$\therefore \neg P$ MT

Suppose that $P \to Q$ is True but $\neg Q$ is also True. Then, for $P \to Q$ to be True, P must be False because if P were True, then $P \to Q$ would be True $\to$ False. So, P must be False. And therefore, $\neg P$ must be True. This concludes the justification of MT. All other rules can be similarly justified. Thus, the system of As, MP, CP, and MT (and RAA) is sound.

One disturbing element that slightly complicates the rosy picture presented above is the fact that some premises cannot possibly be True—they are contradictory. Assuming that we add the connective $\leftrightarrow$, take $P \leftrightarrow \neg P$ as a premise. Any conclusion logically follows from it simply because it's impossible to make $P \leftrightarrow \neg P$ True. Thus, it's impossible for both $P \leftrightarrow \neg P$ to be True and anything else False. This type of argument has a fancy Latin name: *ex falso quodlibet*. (Anything follows from a contradiction.) Structurally, the argument looks like this:

$P \leftrightarrow \neg P$
$\therefore Q$

Well, there's actually one other disturbing element: a conclusion that cannot be False follows validly in an (English) argument and also follows validly in a symbolic proof. For instance, consider $P \leftrightarrow P$ as the conclusion of a symbolic argument. Since it's impossible to make $P \leftrightarrow P$ False, it's impossible to have True premises and a False conclusion. The Latin name for this is *verum sequitur ex quodlibet*. (Truth follows from anything.) Structurally, that argument looks like this:

Q
$\therefore P \leftrightarrow P$

There are different branches of logic that reject the two unusual inferences, *ex falso quodlibet* and *verum sequitur ex quodlibet*. But we won't explore them here. Instead, this seems to be a good place to explain "logically true" and "logically following from."

First, recall the atomic sentences of sentential logic, which we presented at the very beginning of this chapter:

$P, Q, R, S, P_1, Q_1, R_1, S_1, \ldots P_{325}, Q_{325}, R_{325}, S_{325}, \ldots P_{1000}, \ldots$

Also recall how more general sentences are built up from atomic sentences with the addition of connectives. Now, look at a sentence in general. Call it φ. If φ is True no matter what truth assignments are made to its component atomic sentences, then φ is logically True. For example, $P \lor \neg P$ is *logically true*.

More generally, take a set of sentences—call them Γ ("gamma")—and a single sentence φ. φ logically follows from Γ provided that no matter what truth values are assigned to the atomic sentences in Γ, φ comes out True. Consider the set of sentences $\{P, P \to Q\}$. Q *logically follows* from this set, which establishes that the inference of Q from P and $P \to Q$ (i.e., MP) is valid.

A set of sentences Γ is "consistent" if for no sentence ψ and its negation ¬ψ is it possible to prove them both from the sentences in Γ. To put it a little more simply, Γ is consistent if there is some sentence (any sentence) φ such that φ is *not* provable from Γ. We write that φ is not provable from Γ like this:

$$\Gamma \nvdash \varphi \qquad \qquad \text{"Gamma does not prove phi."}$$

If $\Gamma \nvdash \varphi$, then adding the premise ¬φ to Γ does not change its consistency. If φ is not provable from Γ, then Γ is consistent with or without the inclusion of ¬φ in its constituent sentences. We can write this with the symbol ∪ for union or elements belonging to one or multiple sets:

$$\Gamma \cup \{\neg\varphi\} \text{ is consistent.} \qquad \text{"The inclusion of ¬φ in Γ is consistent."}$$

Metalogic—Strong Completeness and Compactness

We now wish to outline a nifty proof that sentential logic is "strongly complete." This means that if you take any set of sentences, call them Γ, and any sentence φ, if φ follows logically from Γ, then there is a proof of φ from Γ. We can write this in symbols but need to first introduce a few new ones:

$\vDash$ "follows from"
$\vdash$ "proves"

We can now write the above definition of strong completeness as follows:

$$(\Gamma \vDash \varphi) \to (\Gamma \vdash \varphi) \qquad \text{"If phi follows from gamma, then gamma proves phi."}$$

We proceed by considering its contrapositive, which we arrive at by flipping the antecedent and consequent and negating them both—i.e., the contrapositive of the statement "if *A*, then *B*" is "if not *B*, then not *A*."

$$(\Gamma \nvdash \varphi) \to (\Gamma \nvDash \varphi) \qquad \text{"If gamma does not prove phi, then phi does not follow from gamma."}$$

Assume that $\Gamma \nvdash \varphi$, meaning that Γ must be consistent. We now state (but do not prove) a simplified version of some fairly complicated logic. It's something you may wish to look up but don't need to know more about here: Lindenbaum's lemma, which yields a simplification of Leon Henkin's completeness proof. We will introduce two new Greek letters to serve as our variables—Δ ("delta") and χ ("chi")—but you shouldn't worry about those, either.

According to our simplified version of Lindenbaum's lemma, from any consistent set of sentences Δ_0, there is a set Δ that includes Δ_0 and contains *every* atomic sentence (sentence letter) χ such that $\Delta_0 \cup \{\chi\}$ is consistent; that is, such that the inclusion of χ in Δ_0 is consistent. Furthermore, Δ is maximally atomic consistent, which means that no additional atomic sentence can be added to Δ without removing its consistency.

Δ becomes a truth value assignment of the atomic sentences so that all atomic sentences in Δ are assigned the value True (and those not in Δ are False). As a consequence, φ becomes False. Thus,

$\quad \Delta \nvDash \varphi$ "Phi does not follow from delta."

And since Γ is a subset of Δ according to Lindenbaum's lemma,

$\quad \Gamma \nvDash \varphi$ "Phi does not follow from gamma."

This completes the extremely rough sketch of the proof. I hope that despite all of the symbols, the underlying logic is obvious.

You may have noticed that we've shifted from merely doing logic to talking about logic. This is the domain of "metalogic." We've claimed that both soundness and (strong) completeness hold for sentential logic, and we've sketched the proofs. Perhaps trying the minds of the less technically proficient, I now would like to show you that strong completeness proves another interesting metatheorem: compactness.

Compactness means that if a sentence φ follows logically from some infinite set Γ, then φ follows from some finite subset of Γ, say Γ'. This is apparent, since "following logically" from an infinite set means (by strong completeness) that there's a finite proof of φ from Γ.

The reader who is aware of a much richer language than sentential logic, namely first-order logic, may wonder whether a similar simplification of Henkin's completeness proof can be used to prove strong completeness for first-order logic. My personal answer is a *resounding* "no," but I leave it an open question whether with some formal modifications it can be accomplished.

Chapter 19
Functions

Cardinality

The size of a set, called its cardinality, is determined by the number of its elements. In the theory of sets, there are two major types of cardinalities: those that are finite and those that are infinite. There are infinitely many finite cardinalities. And we'll see that there are infinitely many infinite cardinalities as well.

For a set with finite cardinality—i.e., a finite set—the obvious way to measure its cardinality is to match each element in the set with a natural number, starting with 1. If we match the natural numbers from 1 to 20 to the elements of a given set, then the set itself has cardinality 20. This is unsurprising, but we'll soon see that the matching method has far-reaching consequences.

Before continuing, however, we should explain the notion of matching more precisely. There are two conditions for matching elements of one set to those of another set. The first condition is this: no two elements of the first set are assigned to the same element in the second set. The second condition: each element in the second set has an element of the first set assigned to it. Condition number one makes the matching a "1-1 matching." And condition number two makes the matching an "onto matching." So, when a set has, for example, 20 elements, that means that 20 numbers (from the set of numbers) have been assigned 1-1 and onto the given set. The more technical word for matching is "function." A 1-1 and onto matching is a 1-1 and onto function. And when there's a function from set A to set B, the elements in A are called the "domain" of the function (Dom), and those in B are called the function's "range."

Now, let us see what happens when we match guests (1-1 and onto) to rooms in an infinite hotel.

Hilbert's Hotel

"Hilbert's hotel," named after the great mathematician David Hilbert (Fig. 15.7), is an imaginary hotel with infinitely many rooms numbered 1, 2, 3, and so on. Hilbert supposedly invented the hotel to explain to his students how infinity works.

What happens when Hilbert's hotel is full—there are infinitely many guests—and someone shows up wanting a room? We suppose that the desk clerk, a mathematically sophisticated gent, simply moves each guest to the next room in the hotel. Guest 1, who is presently in Room 1, is moved to Room 2. Guest 2 is moved to Room 3. And so on. Now Room 1 is free for occupancy, as shown in the diagram in Fig. 19.1.

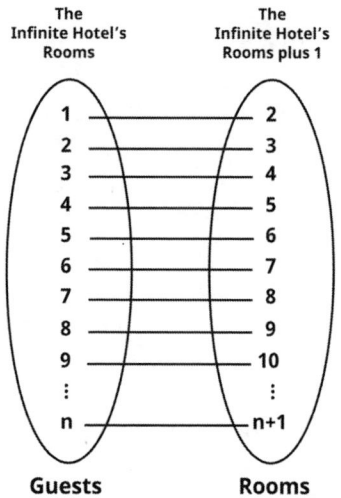

Figure 19.1 Diagram of Hilbert's hotel problem: $n + 1$. Room 1 is now free for occupancy. Image credit: Ernesto Mora.

So, the new guest is ushered into Room 1. But what if infinitely many persons show up wanting a room?

In this case, the clever desk clerk moves every guest in room n to room $2n$. That is, the guest in Room 1 is moved to Room 2, the guest in Room 2 is moved to Room 4, the guest in Room 3 is moved to Room 6, and so on. What this does, as Fig. 19.2 shows, is free up all the odd-numbered rooms.

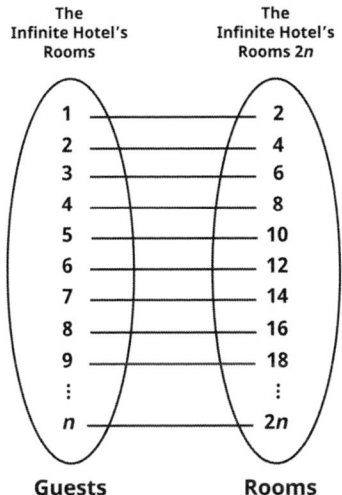

Figure 19.2 Diagram of Hilbert's hotel: 2*n*. All odd-numbered rooms are empty. Image credit: Ernesto Mora.

The infinitely many new guests settle into the odd-numbered rooms, and the hotel is full once more.

Hilbert's hotel illustrates the first infinite cardinality, a set with denumerably many elements, those that can be matched (1-1 and onto) to the natural numbers. We've seen, though, that paradoxically the natural numbers can be paired with the even natural numbers, which comprise only half of the naturals. That's just one seeming paradox that shows up with infinity.

Here's another seeming paradox. In this one, each guest is numbered sequentially, beginning with 1. Then, the room assigned to that guest is the square of the guest's number. So, Guest 1 is assigned to Room 1, Guest 2 is assigned to Room 4, Guest 3 to Room 9, and so on, as depicted in Fig. 19.3.

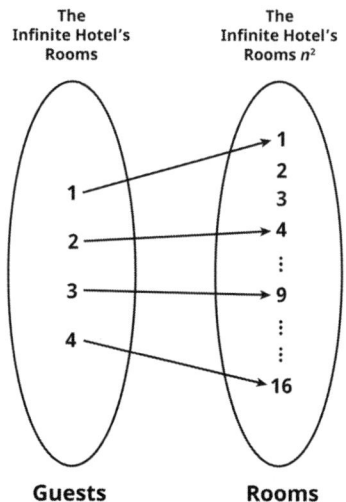

Figure 19.3   Diagram of Hilbert's hotel: n^2. Image credit: Ernesto Mora.

So, this list of squares is denumerable as well.

If the even numbers, the odd numbers, the even-and-odd numbers, and the squares of numbers are all denumerable, where does the first non-denumerable number show up? For instance, are the fractions non-denumerable? If they are not denumerable, we need to show that there's no way to match all "rationals" (numbers that can be expressed as the fraction of two integers) with the "naturals" (positive integers). Some rationals must go unmatched.

On the other hand, if the rationals are denumerable, we must illustrate a way of matching them to the naturals. It turns out that they actually are denumerable, and here's how the matching works. Follow the arrows in the chart in Fig. 19.4:

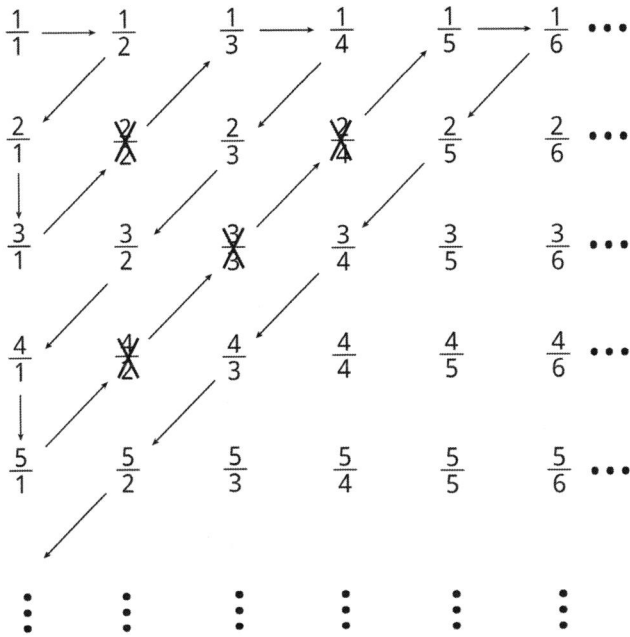

Figure 19.4 Diagram of Hilbert's hotel: rationals. Image credit: Ernesto Mora.

Using the rooms in Hilbert's hotel, Fig. 19.4 shows that the guest numbered 1/1 can be placed in Room 1, guest numbered 1/2 in Room 2, Guest 2/1 in Room 3, and Guest 3/1 in Room 4. Then, notice something odd: Room 2/2 is crossed out. This is because the guest whose fractional number is 2/2 is the same as Guest 1/1, since $2/2 = 1$. But we don't need to eliminate equal fractions; we can include all infinitely many of them too. Just give Guest 2/2 a unique new room, and then ignore the other crossed out fractions. That way, 2/2 goes to the next natural number, Room 5.

Exercise 40: A simple computer program could be written showing which fraction is paired with which whole number. See whether you can write such a program to show this.

As you can see, infinitely many fractional-numbered persons (supposing that's possible) can find rooms in Hilbert's hotel. More technically, we have shown that there's a 1-1 and onto function from the natural numbers to the rationals. Another way of saying this is that there is a bijection from the naturals to the rationals.

Georg Cantor

Figure 19.5 Photograph of Georg Cantor (1845–1918). (Author unknown/Wikimedia Commons.)

It may begin to appear that there's just a single infinity, that all infinite sets are denumerable. That this is not so was shown by a groundbreaking proof from the creator of set theory, Georg Cantor. Cantor's diagonal argument shows that there are more irrational numbers than natural numbers. Irrational numbers are numbers that *cannot* be expressed as the fraction of two integers —e.g., $\sqrt{2}$ or π.

We'll take just the irrationals between 0 and 1 and show that there are more of them than naturals. So, if the arriving guests of Hilbert's hotel can be numbered by the irrationals, even the cleverest desk clerk cannot secure them rooms.

We begin by supposing that we can list the irrational numbers between 0 and 1 just like we can list the natural numbers—i.e., 1, 2, 3, and so on.

Suppose that the list of irrationals begins like this:

    0.4370152847...
    0.5916683243...
    0.8163074502...

0.0951038726...
0.1089037243...
0.0788423125...

...

Cantor's diagonal argument is so called because we take the diagonal of the column of numbers—the first place in the first column, the second place in the second column, the third place in the third column, etc.—and add 1 to every number in the diagonal unless it's 9, in which case we substitute 8. To illustrate, the diagonal of these numbers is marked in **bold** and its substitution is provided after the ellipses:

0.**4**370152847... 5
0.5**9**16683243... 8
0.81**6**3074502... 7
0.095**1**038726... 2
0.1089**0**37243... 1
0.07884**2**3125... 3

...

We arrive at the new number: 0.587213..., which is different from any number on the list and hence cannot itself be on the list. Why is it different? Well, the first digit after the decimal point of the new number, 0.587213..., is a 5. But on the list itself the first digit of the first number after the decimal point is a 4. So, the new number differs from the first number on the list in the first position (after the decimal point). Now, take the second digit (after the decimal point) of the new number, 0.587213... It's an 8. But on the list, the second digit of the second number is 9. The third number of the new number, 0.587213..., is a 7, which differs from the third digit of the third number on the list, 6. Continuing in this way, we can see that the nth digit of the new number differs from the nth digit of the nth number in the original list.

Objecting to Cantor's proof, opposing his mathematical theories, and disliking Cantor's methods in general, Leopold Kronecker banished all but the positive integers from mathematics. He is reputed to have said, "The good Lord made the integers; everything else is the work of man."[22] Contrary to much nasty writing to the effect that Kronecker and Cantor were bitter enemies, their relations were in fact quite amicable. They even shared a dinner table on occasion. It was only of Cantor's mathematics that Kronecker disapproved.

If we accept Cantor's proof, we find that there cannot be a 1-1 correspondence between the natural numbers and the "reals," all numbers on a continuous number line with no gaps. Real numbers include all naturals, integers, rationals, and irrationals. *There are more real numbers than natural numbers.* Cantor demonstrated this by showing that there are more irrational numbers between 0 and 1 than natural numbers. Another way to put this is that there is no 1-1 and onto function from the natural numbers to the real numbers.

Exercise 41: Consider binary decimals, represented by 1 and 0 after the decimal point. For instance, one binary decimal is 0.000..., another is 0.111..., and a third is 0.101010... Prove that an infinite binary decimal series is denumerable, or that it's not.

In his book *Gödel, Escher, Bach,*[23] Douglas Hofstadter found a witty analogy to Cantor's diagonal argument. It begins with the following list of mathematicians:

De Morgan
Abel
Boole
Brouwer
Sierpinski
Weierstrass

Next, we **bold** the diagonal letters of these names, like so:

De Morgan
A**b**el
Bo**o**le
Bro**u**wer
Sierp**i**nski
Weiers**t**rass

Then, we take the letter that alphabetically precedes each darkened diagonal letter. This spells out **Cantor**. Clever, isn't it?

Now we wish to examine Cantor's theorem, which states that the power set of any given set has greater cardinality than the set itself. We could express this same idea as follows: there's no onto function from set S to $\mathscr{P}(S)$. $\mathscr{P}(S)$ here represents the power set of S: the set of all subsets of S.

Before we start, I need to introduce two more symbols:

$\in$	"is an element of"
$\notin$	"is not an element of"

If we wanted to show that some set S contains an element x, we would write: $x \in S$. Onto the proof!

Let f be any function from S to $\mathscr{P}(S)$. Now, specify a subset B of S, an element of $\mathscr{P}(S)$ such that

$$B = \{x: x \in S \text{ and } x \notin f(x)\}$$

B must then differ from every element f assigns to $\mathscr{P}(S)$ (by at least a single element). But f is taken to be any arbitrary function from S to $\mathscr{P}(S)$. So, the power set of any set must have greater cardinality than the set itself.

We will now define a second sort of matching function. Recall that the first sort is a 1-1 and onto matching—a bijection. The second type of function, match_2, will be called a "1-1 both ways" matching, an injection in both directions. To match_2 the elements of A and B, each element of A must be assigned to an element of B and each element of B must be assigned to some element of A. The two types of matching, match_1 and match_2, yield the exact same result. That is, they are equivalent. However, they differ conceptually, and thus must be proven equivalent.

What we wish to prove is $\text{matching}_1 \rightarrow \text{matching}_2$. Spelled out in English, we want to show that if there's a 1-1 and onto function from set X to set Y, then there's a 1-1 function from set X to set Y *and* a 1-1 function from set Y to set X.

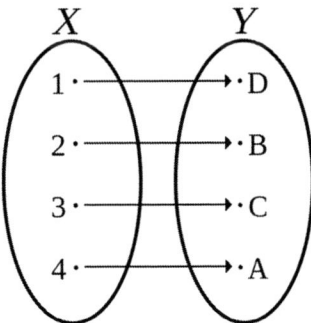

Figure 19.6 Sample diagram of a matching function (bijection) $f\colon X \rightarrow Y$. Image credit: Schapel / Wikimedia Commons.

In Fig. 19.6, we let f be the 1-1 and onto function that matches numbers to uppercase letters: $f(1) = D$, $f(2) = B$, $f(3) = C$, and $f(4) = A$. The function can also be represented by this set of ordered pairs: $f = \{(1, D), (2, B), (3, C), (4, A)\}$.

Taking our cue from the diagram, to obtain the needed 1-1 matching function Y to X, we simply reverse f. This reversal is called f inverse, and it's written

f^{-1}. $f^{-1} = \{(D, 1), (B, 2), (C, 3), (A, 4)\}$. f can be reversed, since f is onto, meaning that every element of Y (the range of Y) has been assigned a value by the function f. The f^{-1} function of any element y in Y simply returns the x in X such that $f(x) = y$. No matter how many elements are in sets X and Y, this inverse exists—and that's all there is to the proof.

Now that we've proved this simple theorem, we turn to its converse, which is not at all simple. Namely, matching$_2$ → matching$_1$.

Proof of this direction, in contrast to the earlier one, is a genuine theorem used often by set theorists. The earlier proof was just invented for comparison. The hard proof is alternatively called the Schröder–Bernstein theorem, the Cantor–Schröder–Bernstein theorem, or the Cantor–Bernstein theorem. Here it is, presented in outline form:

To Prove:
If there's a 1-1 function from set X to set Y and a 1-1 function from set Y to set X, then there's a 1-1 and onto function from X to Y (a 1-1 correspondence from X to Y).

Outline of Proof:
First, without loss of generality, make X and Y disjoint.
Now, let f be a 1-1 function from X to Y and let g be a 1-1 function from Y to X.
Each element x in X has infinitely many *descendants*, $f(x)$, $g(f(x))$, $f(g(f(x)))$, ….
Each element y in Y has infinitely many *descendants*, $g(y)$, $f(g(y))$, $g(f(g(y)))$, ….
Every term in the sequences is an *ancestor* of each subsequent term, and the immediately preceding term is a *parent* of its child.
Tracing the ancestry of any term in either sequence leads to an element of X that has no parent—i.e., an *orphan* $(X - g(y))$, or an *orphan* $(Y - f(x))$ —or else the parentage continues without end.
Call the three sets that originate in X, respectively, X_x, X_y, and X_∞. Call those that originate in Y, respectively, Y_y, Y_x, and Y_∞.
If $x \in X_x$, then $f(x) \in Y_x$, and the restriction of f to X_x is a 1-1 correspondence between X_x and Y_x;
If $x \in X_y$, then $x \in$ Dom of g^{-1} and $g^{-1}(x) \in Y_y$. The restriction of g^{-1} to X_y is a 1-1 correspondence between X_y and Y_y;
If $x \in X_\infty$, then $f(x) \in Y_\infty$ and the restriction of f to X_∞ is a 1-1 correspondence between X_∞ and Y_∞.
Weaving these three correspondences together gives us a 1-1 and onto matching between X and Y.

There are several alternative proofs of this important theorem, many accompanied by appealing diagrams. Each diagram for a given proof illustrates how to weave together two distinct 1-1 functions to compose a single 1-1 and onto function.

Although these quite distinct diagrams seem clearly to point to formally distinct proofs, some mathematical foundationalists have claimed that there's essentially only a single unique proof of this theorem (without the axiom of choice). Obviously, this claim relies on some special meaning of the word "essentially." But its analysis lies beyond the scope of this book. If you're interested, there's an entire book on this subject, conveniently entitled *Proofs of the Cantor–Bernstein Theorem: A Mathematical Excursion* by Arie Hinkis.

This raises the question of which of the two kinds of matching is the right kind to determine cardinality. The answer is that the bijection is the foundationally significant one. The other sort of matching, the 1-1 both ways matching, is ad hoc and lacks the same intuitive force. So, we can say that the cardinality of a set is determined by the sets that match it 1-1 and onto: all sets for which there exists a bijection.

We know that the set of real numbers is greater than the set of naturals, that there is no bijection between them. The cardinality of natural numbers is typically written as $\aleph_0$ ("aleph null"). The set of reals, which has cardinality $2^{\aleph_0}$, is larger than $\aleph_0$. But how much larger? The infinite cardinalities line up like this: $\aleph_0$, $\aleph_1$, $\aleph_2$, ... So, we can ask whether $2^{\aleph_0} = \aleph_1$. Do the reals constitute the next infinity after the naturals? This question is labeled the "continuum hypothesis." In 1940, Kurt Gödel showed that the continuum hypothesis can't be proven false from the axioms of set theory. And in 1963, Paul Cohen proved that the continuum hypothesis can't be proven true. Thus, the continuum hypothesis holds a special place in set theory: it is independent of set theoretic axioms.

Who Invented Cantor's Back-and-Forth Argument?

It's a strange fact in mathematics that many of the developments named for one famous figure can be attributed to another. One reads that Peano's postulates are really Dedekind's, that Dedekind's chains are really Frege's, and that Newton's method was known to Archimedes. So, we raise the question: Who really invented Cantor's back-and-forth argument, which we'll abbreviate BAF?

It is a well-entrenched belief among set theorists and model theorists that BAF comes from Cantor, who first introduced it to prove the isomorphism of any two countable, dense linear orders (with or without endpoints). But in 1975, I discovered a proof of the isomorphism theorem that did not require the BAF technique. Since this simplification was surprising, I pursued the matter a bit at the time, which led me to look up Cantor's famous proof. I was shocked to discover that Cantor did the proof *my way*, not using the BAF technique that has been associated with him. Now I had two surprising facts to consider. First, that the isomorphism theorem did not require BAF. Second, that Cantor did not use BAF himself, at least not where he supposedly introduced it.

The resolution of the matter had to wait until 1993, when a piece by Jeremy Plotkin entitled "Who Put the 'Back' in Back-and-Forth" in the book *Logical Methods* noted that E. V. Huntington in 1905 used a back-and-forth argument and "by failing to say otherwise credited Cantor with its invention."[24]

The Back-and-Forth Argument (BAF)

The purpose of BAF is to create a shoelace effect between two sets in order to tie them together in a 1-1 and onto map. The idea is to define two functions, f and f', such that f goes "forth" from X to Y (i.e., $f: X \to Y$) and f' goes "back" from Y to X (i.e., $f': Y \to X$). Piecing them together creates a 1-1 correspondence h between X and Y. If, in addition, h preserves the order relations on each of the two sets, then h is an isomorphism.

Let X and Y be any two denumerable, dense linear orders without endpoints. The elements in the two sets can be arranged in two lists, where the order of the elements in the lists is indicated by the number of prime symbols appended to the letter x or y, such that the first element in the list has one prime, the next one has two primes, etc:

$$X = \{x', x'', x''', x'''', \ldots\}$$
$$Y = \{y', y'', y''', y'''', \ldots\}$$

Start out by letting $x_0 = x'$ and $y_0 = y'$. Then go forth and take the next y in the list, which is y''. y'' now becomes y_1, which is greater than or less than y_0. Suppose that y_1 is less than y_0. Then choose an x less than x_0 and make this x_1. (If y_1 is greater than y_0, choose an x greater than x_0 to be x_1.)

Now, go back to X and take the first element left in

$$\{x', x'', x''', x'''', \ldots\} - \{x_0, x_1\}$$

We know it cannot be x' because x' is x_0, but it could be x''. Let the first element in this set be x_2. x_2 is either greater than each of x_0 and x_1, less than both of them, or between them. Now we need to do the same thing with y:

$$\{y', y'', y''', y'''', \ldots\} - \{y_0, y_1\}$$

Choose a y that bears the same relation to y_0 and y_1 as x_2 does to x_0 and x_1. This y then becomes y_2.

Go forth again to Y. Get the next y not yet taken. Call this y_3. Then choose an x such that it bears the same relation to the three x's already taken as y does to the three prior y's.

This method constitutes a recursive definition of the two sequences $<x_0, x_1, x_2, x_3, \ldots>$ and $<y_0, y_1, y_2, y_3, \ldots>$ such that the function h is defined from X to Y, where $h(x_n) = y_n$ is the isomorphism we wanted.

The One-Way Argument

The one-way argument creating an isomorphism is similar to the above construction in one direction, except that we explicitly take *the first* value satisfying the given property.

We define $g\colon X \to Y$ as follows, and then prove the onto part. Let $g(x') = y'$ just to fix the initial points. Then, if x'' is less than x', let $g(x'') =$ the first y such that y is less than y'. Or, if x'' is greater than x', let $g(x'') =$ the first y such that y is greater than y'. For any x^n (i.e., x followed by n primes), $g(x^n) =$ the first y among

$$\{y', y'', y''', \ldots, y''''\} - \{g(x'), g(x''), \ldots, g(x^{n-1})\},$$

such that y bears the same relation to all of $g(x'), g(x''), \ldots, g(x^{n-1})$ as x^n bears to all of $x', x'', \ldots, x^{n-1}$.

The claim is that g is an isomorphism, as h was in the BAF argument. It was easier to see that h is an isomorphism because the back-and-forth character of the construction ensured that X and Y would be "used up together"—i.e., that h is an onto function, also called a surjection.

Suppose that at least one element of Y is missed. Take the first such element y^m, such that y^m is *not* the image of any x in X under g. This means that all of $y', y'', \ldots, y^{m-1}$ are images of x's under g. Suppose that it takes k-many x's to cover these $m - 1$ elements of Y, where k is at least as large as $m - 1$. That creates $k + 1$ "slots," one of which y^m fits into. For example, one slot exists between the next-to-largest and the largest element among $g(x')$, $g(x'')$, ..., $g(x^k)$, and another exists past the largest element (since there's no endpoint). Not only does y^m fit into one of these slots, but it is the first such available element that does. For example, suppose that y^m is between $g(x')$ and $g(x'')$. Then, as soon as an x is reached that is between x' and x'', g of that x will equal y^m. At some point, an x will be encountered that fits into that slot (by denseness), and it will be mapped to y^m. Therefore, g is surjective, which is what we wanted to show.

Chapter 20
Set Theory

Before looking at the axioms of modern set theory, let's glance at the set theory of Gottlob Frege.

In Frege's set theory, a well-defined property could determine a set. For instance, take the property of being a desk. We should be able to define a set of all things that have the property of being desks. More concisely, if D is the property of being a desk, then there should be a set of all desks, $S = \{x: Dx\}$. The problem with Frege's theory of sets was his unrestricted comprehension axiom.

Frege's unrestricted comprehension axiom states that to every condition—e.g., being a desk—there corresponds a set of things meeting the condition. We could write this in logical notation as follows:

$$\exists y \forall x (x \in y \leftrightarrow \varphi x)$$

We define φx as "any property of (only) x." Accordingly, this sentence says, "There exists a y such that for all x, x belongs to y if and only if it is any property of (only) x."

Russell's Paradox

British philosopher and mathematician Bertrand Russell discovered a paradox that turns Frege's set theory on its head. Russell supposed the property of x to be "x does not belong to itself," written as $x \notin x$. So, Russell's instance of comprehension in Frege's set theory is

$$\exists y \forall x (x \in y \leftrightarrow x \notin x)$$

We could write this in plain English as "... x belongs to y if and only if x doesn't belong to itself." From this, a contradiction follows. You can see this below:

$\exists y \forall x (x \in y \leftrightarrow x \notin x)$

Let y be the set A; we then have

$\forall x (x \in A \leftrightarrow x \notin x)$

Since this is true for *any* set, let this set also be A. Thus,

$A \in A \leftrightarrow A \notin A$ ... a contradiction!

Russell's paradox destroyed the entire edifice of Gottlob Frege's set theory, although its demolition spawned several different sorts of set theories.

Russell wrote a letter to Frege, which has since become famous, informing him that the set of things not belonging to themselves leads to a contradiction in his set theory. Frege was about to release the second volume of his magnum opus when he received this letter from Russell.

Gottlob Friedrich Ludwig Frege
Prof. d. Mathematik
(* 1848. 1874. + 1925)

Figure 20.1 Photograph of Gottlob Frege (1948–1925). Photo credit: Friedrich-Schiller-Universität Jena.

Figure 20.2 Photograph of Bertrand Russell (1872–1970). Photo credit: Yousuf Karsh for Anefo / Wikimedia Commons.

Frege acknowledged Russell's work in an appendix to his second volume. He wrote: "Hardly anything more unwelcome can befall a scientific writer than that one of the foundations of his edifice be shaken after the work is finished. This is the position into which I was put by a letter from Mr. Bertrand Russell as the printing of this volume was nearing completion."[25]

Later, Russell wrote about Frege:[26]

> As I think about acts of integrity and grace, I realize that there is nothing in my knowledge to compare with Frege's dedication to truth. His entire life's work was on the verge of completion, much of his work had been ignored to the benefit of men infinitely less capable, his second volume was about to be published, and upon finding that his fundamental assumption was in error, he responded with intellectual pleasure clearly submerging any feelings of personal disappointment. It was almost superhuman and a telling indication of that of which men are capable if their dedication is to creative work and knowledge instead of cruder efforts to dominate and be known.

Before abandoning Russell's paradox, let's look at a possible example in the real world. Imagine a club with a single rule of membership: applicants must pass rule A to be admitted. But rule A itself states that the applicant must fail rule A. So, anyone passing rule A would consequently be rejected by passing rule A. And anyone failing rule A would then, by rule A, be accepted for not passing.

Groucho Marx has a humorous take on this paradox. He resigned from a club, saying he would not belong to any club that would accept him as a member. He also said to anti-Semitic members of a swim club who had refused admission to his daughter: "She's only half Jewish. How about if she only goes in up to her waist?"[27]

Russell's paradox in set theory has become very famous. But there's a more trivial version of it that Russell, who churned out the numerous theorems in his master work *Principia Mathematica*, should have already known. To introduce a simpler but informative version of Russell's paradox in first-order logic, we must first take a brief excursion into first-order logic—but just enough to get the basic idea.

As in set theory, "∃" stands for "some" and "∀" stands for "all." When $\exists y(Fy)$ —"something has the property F"—is on a line in a derivation, you are permitted to existentially instantiate y (take an instance of y) to y_0 only if y_0 hasn't previously been used in the derivation. When $\forall x(Hx)$ appears on a line, you can universally instantiate x to any letter you wish.

The letters "*Gxy*" indicate a relationship between two things, *y* and *x*. More generally, "$\exists y \forall x \, Gxy$" just says that for some *y*, every *x* is related by *G* to *y*. For example, for some person *y*, every person *x* lives in the same city that *y* does. This example is clearly false—who could this person be?

Here's a counterpart to Russell's paradox in first-order logic that mirrors the proof of paradox in Frege's set theory.

1. $\exists y \forall x (Gxy \leftrightarrow \neg Gxx)$ Assumption (which leads to a contradiction)
2. $\forall x (Gxy_0 \leftrightarrow \neg Gxx)$ Existential Instantiation: $y \rightarrow y_0$
3. $Gy_0 y_0 \leftrightarrow \neg Gy_0 y_0$ Universal Instantiation: $x \rightarrow y_0$

    ... A contradiction!

In light of the inconsistency in Frege's axioms of set theory, modern set theory has revised axioms that seem free of contradiction. Instead of presenting the axioms of set theory in all their technical glory, we will first look at a conception of set theory from which almost all of the axioms seem to follow.

The Iterative Conception

Although there were similar conceptions earlier, the phrase "the iterative conception of set" comes from the title of a 1971 article of that name by George Boolos. Boolos writes that the failure of Frege's set theory might make it seem that any alternative view of sets designed to avoid the paradoxes would be arbitrary, having no greater intuitive basis than any other. This view, he tells us, does not take into consideration the iterative conception.[28]

We start out with the simplest set of all, the empty set, or, since we don't want to use the word "set"—that's a technical term—we'll call it the empty collection. The empty collection is intuitive. It's what's in your pocket when it's empty. It's what you've collected in the bank when you've run out of money. Call it ∅.

Starting with ∅, we'll build up bigger and bigger collections. First, we must discuss "levels." (Boolos calls them "stages."[28]) We'll say that ∅ is at the bottom level; just ∅ and nothing more. Using braces, ∅ = { }, the single collection with nothing in it. We now need a way to generate more collections from this one. What we'll do at the next level is to take { } plus all collections of { }. This gives us two collections, { } itself and the collection of that collection, {{ }}. We began with a single collection and now we have two. On the next level are three collections: { }, {{ }}, {{{ }}}. It's clear that if we continue in this way, we'll have *n* collections on level *n*, the largest of these collections having *n* braces around it. If we wish, we can identify the numbers with collections, letting { } stand for the number 0, meaning that 0 is the number of elements in the earliest collection (or 1 is the number of braces).

What else? Well, in order to arrive at a basis for all of modern set theory—and most of the axioms stated above—we must continue building up collections past all finite levels. After all finite levels—0, 1, 2, ..., n—a new level must be created, an infinite level called level omega, written like this: ω. ω collects all earlier collections, { }, {{ }}, {{{ }}}, ... And (as if that weren't enough), we still keep going: to level $\omega + 1$, $\omega + 2$, $\omega + 3$, ..., 2ω, and on, and on, and on.

Many things could be said of the iterative conception as described above. But we'll mention only a few. Remember, the claim was made that the conception of set is intuitive. Maybe it is, maybe it's not—it all depends on whose intuitions we're considering. One possibly *unintuitive* feature is that there's a level beyond all the finite ones, level ω. Before we let this feature of "beyondness" incline us to hastily abandon this conception, we must keep in mind that virtually all mathematicians these days accept the notion of infinity. Even more than that, they accept that there is not just a single infinity. We've proved that beyond the infinity of natural numbers is a further infinity of real numbers. That is, taking $\aleph_0$ as the cardinality of the natural numbers, there's a further cardinality c of the continuum of the real numbers. And Cantor proved that there are infinitely more.

Axiom of Sets

The axioms listed below are presented in the raw, so to speak. With a few exceptions, all of the symbols used below should be familiar by now. (In the power set axiom, "$\subseteq$" means "is a subset of"; in the replacement axiom, "$\exists!y$" means "there exists a unique y"; in the axiom of choice, "f" means "a function.")

Extensionality Axiom: Two sets are equal if and only if they have the same elements.

$$\forall x(x \in A \leftrightarrow x \in B) \rightarrow A = B$$

Subset Axiom (schema): There's a set B whose elements are those in set A with property φ.

$$\exists B \forall x(x \in B \leftrightarrow (x \in A \land \varphi x))$$

Pairing Axiom: There's a set A containing just elements y and z.

$$\exists A \forall x(x \in A \leftrightarrow x = y \lor x = z)$$

Sum Axiom: There's a set C whose elements are those in B where B

belongs to A.

$$\exists C \forall x (x \in C \leftrightarrow \exists B (x \in B \wedge B \in A))$$

Power Set Axiom: There's a set B consisting of all subsets of A.

$$\exists B \forall C (C \in B \leftrightarrow C \subseteq A)$$

Regularity Axiom: For non-empty set A, there's an element x of A where y belongs to x if and only if y doesn't belong to A.

$$A \neq \varnothing \rightarrow \exists x [x \in A \wedge \forall y (y \in x \leftrightarrow y \notin A)]$$

Replacement Axiom: If the domain of function φ is a set, so is its range.

$$(\forall x)(\exists ! y)(\varphi(x, y)) \rightarrow (\forall X)(\exists Y)(\forall z)(z \in Y \leftrightarrow (\exists w \in X) \wedge \varphi(w, z))$$

Axiom of Choice: For any set x of non-empty disjoint sets, there's a set of a single element from each of the disjoint sets.

$$\forall x (\forall y \in x \; y \neq \varnothing \rightarrow \exists f \; \forall y \in x \; f(y) \in y)$$

Axiom of Infinity: In brief, there's an infinite set.

$$\exists x (\varnothing \in x \wedge \forall y (y \in x \rightarrow y \cup \{y\} \in x))$$

The above are (somewhat redundant) axioms of Zermelo–Fraenkel set theory together with the axiom of choice. Almost all of them follow from the "level" theory of the iterative conception. However, extensionality doesn't follow; it just registers our intention to call sets equal when they have exactly the same elements. The axiom of choice doesn't strictly follow either, though one could ask rhetorically, "how could the choice sets be *missing*?" After all, at each level, we said to take all collections. And the infinite sets exist only because we said that there were infinite levels after all the finite ones. Had we said to stop right there, no infinite collections would have been created.

This raises the question whether Boolos' iterative conception of sets is the most intuitive conception of set theory or whether there are others that are more intuitive.

One tempting alternative is Joseph Shoenfield's visualization principle, expressed in his book *Mathematical Logic*. According to this view, sets are built up in "stages," which we call "levels." A level exists wherever possible. For instance, if there is a collection of levels $L_1, L_2, L_3, \ldots, L_n, \ldots$, then we can visualize a further level L_ω.[29] This seems to be a niftier way to get infinite sets than the iterative conception, but is "visualization" simply too vague for such use?

Chapter 21
Group Theory and Other Elementary Theories

Group theory may be considered the most fundamental theory in mathematics, involved in numerous types of investigations. The origins of group theory are often attributed to Évariste Galois.

Figure 21.1   Portrait of Évariste Galois (1811–1832) at about age 15. (Drawing on gray paper by an unknown artist / photograph by Shokichi Iyanaga / Wikimedia Commons.)

A romantic figure, Galois was an ardent, radical French republican who was killed in a duel before his 21st birthday. His short life was mostly filled with disappointments, as he failed entrance examinations into the prestigious École

Polytechnique twice. But on his own at the age of 17, he mastered mathematical works by Legendre and Lagrange and was able to extend their ideas.

Supposedly aware of his imminent doom, Galois, according to legend, furiously scribbled his mathematical ideas the night before a fated duel. It's difficult to assess the accuracy of diverse accounts of Galois' life, as much fancy seems to have been added to the raw facts. But for a particularly elaborate portrait, you could consult E. T. Bell's *Men of Mathematics*.[3]

Galois is credited for coming up with the label "*groupe théorie*." Here are the group theory axioms. (Don't be put off by the unfamiliar symbols! To the extent that you need to understand them, they are explained below.)

G1. $(x \circledast y) \circledast z = x \circledast (y \circledast z)$
G2. $\exists î((x \circledast î = x) \wedge (î \circledast x = x))$
G3. $\forall x \exists x^{-1}((x \circledast x^{-1} = î) \wedge (x^{-1} \circledast x = î))$

G2 says there is an element $î$ such that both $x \circledast î = x$ and $î \circledast x = x$ are true. G3 says that for every element x, there is an element x^{-1} such that both $x \circledast x^{-1} = î$ and $x^{-1} \circledast x = î$ hold.

You can see that if we interpret $\circledast$ to be addition over the set of integers, and interpret $î$ to be 0, then the three axioms hold. $î$ is called the identity element, and x^{-1} is the inverse of x. The identity element for addition is 0, since $x + 0 = x$. And the additive inverse of an element is its negative, which means that adding the negative of a number to its positive yields the identity element, 0.

What happens if we interpret $\circledast$ to be multiplication over the set of integers? First, what's the inverse? Rephrasing the question, what times what number returns the first number? Take 10, for instance. 10 times what equals 10? The number 1, obviously. Now, 10 times its inverse is supposed to deliver 1. Ask yourself, "10 times what equals 1?" Clearly, it must be 1/10. That is, 10 times 1/10 equals 1. But we don't have fractions in our set. Thus, multiplication over the set of integers is not a group. Can we rectify this shortcoming by including all inverses, where the inverse of any number x is $1/x$? No, because we still don't have an inverse for 0. What happens then if we exclude 0? The set of all rational numbers excluding 0 under multiplication does then indeed form a group.

G1 gives us associativity. Notice, however, that we don't have an axiom for commutativity. That is, it is *not* a theorem of group theory that $x \circledast y = y \circledast x$. (We know independently, however, that the two mentioned operations, addition and multiplication, both happen to commute: $x + y = y + x$ and $x \bullet y = y \bullet x$.)

We do have uniqueness for both the identity element and for inverses. Suppose that there were two identity elements, $î_1$ and $î_2$. Then both $î_1 ⊛ î_2 = î_1$ and $î_1 ⊛ î_2 = î_2$. Hence, $î_1 = î_2$. Thus, the identity element is unique. Now suppose that there were two inverses. For instance, let x_1 and x_2 both be inverses of x. Then both $x_1 ⊛ (x ⊛ x_2) = x_1 ⊛ î = x_1$ and $(x_1 ⊛ x) ⊛ x_2 = î ⊛ x_2 = x_2$. But by G1, $x_1 ⊛ (x ⊛ x_2) = (x_1 ⊛ x) ⊛ x_2$. Thus, $x_1 = x_2$ and the inverse is unique.

Consider the equilateral triangle in Fig. 21.2 with vertices 1, 2, and 3.

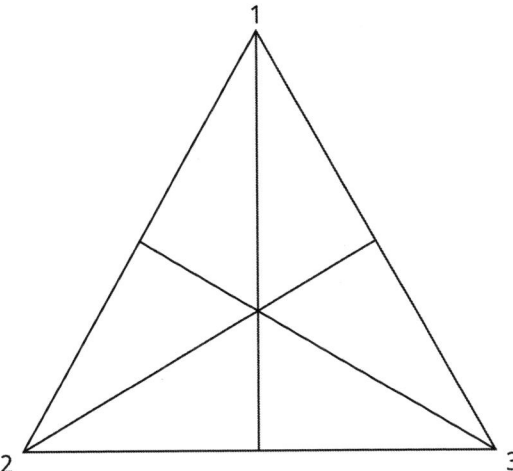

Figure 21.2   Diagram of an equilateral triangle. Image credit: Ernesto Mora.

Let's rotate the triangle 60° clockwise so that 1 is where 3 was, 2 is where 1 was, and 3 is where 2 was. That's one set of transformations. We could rotate the triangle 120°, making 1 go where 2 was, 2 go where 3 was, and 3 go where 1 was. Or we could rotate it 360°, which is the same as not rotating it at all.

Are these three all the transformations there are? No. We could flip the triangle along each of the three drawn lines, giving us three additional transformations, depending on which line we use to flip the triangle. These three additional transformations, plus the three earlier ones, yield the following chart where, for specificity, we first apply the column transformation, then the row. Note that these transformations are identical to the permutations we get when we permute the numbers 1, 2, and 3. We know from simple multiplication that there are six permutations of the numbers 1, 2, and 3; three choices for the first number times two choices for the second and just one choice for the third.

Number	123	213	132	321	231	312
123	123	213	132	321	231	312
213	213	123	231	312	132	321
132	132	312	123	231	321	213
321	321	231	312	123	213	132
231	231	321	213	132	312	123
312	312	132	321	213	123	231

From this chart we can see that these transformations form a group. Permutations are always associative. The group identity is (123) and each element has an inverse. These transformations are not commutative, which is to say they are "non-Abelian." In fact, this set of transformations is the simplest non-Abelian group that there is. Permuting just the letters a and b (i.e., $a \bullet b$) is the same as permuting b and a ($b \bullet a$).

Here are just a few not entirely obvious theorems:

T1. $x \circledast z = y \circledast z \rightarrow x = y$
T2. $x \circledast \hat{\imath} = \hat{\imath} \circledast x$
T3. $y \circledast x = y \rightarrow x = \hat{\imath}$
T4. $x \circledast x = \hat{\imath} \rightarrow x \circledast y = y \circledast x$
T5. $x \circledast x^{-1} = x^{-1} \circledast x$
T6. $(x^{-1})^{-1} = x$
T7. $\hat{\imath}^{-1} = \hat{\imath}$
T8. $(x \circledast y)^{-1} = y^{-1} \circledast x^{-1}$

And here's a proof of T1:

1. $x \circledast z = y \circledast z$	1, As (1)
2. $(x \circledast z) \circledast z^{-1} = (y \circledast z) \circledast z^{-1}$	1, Substitution
3. $(x \circledast z) \circledast z^{-1} = x \circledast (z \circledast z^{-1})$	G1
4. $z \circledast z^{-1} = \hat{\imath}$	G3
5. $(x \circledast z) \circledast z^{-1} = x \circledast \hat{\imath}$	3, 4, Substitution
6. $x \circledast \hat{\imath} = x$	G2
7. $(x \circledast z) \circledast z^{-1} = x$	5, 6, Substitution
8. $(y \circledast z) \circledast z^{-1} = y \circledast (z \circledast z^{-1})$	G1
9. $(y \circledast z) \circledast z^{-1} = y \circledast \hat{\imath}$	4, 8, Substitution
10. $y \circledast \hat{\imath} = y$	G2
11. $(y \circledast z) \circledast z^{-1} = y$	9, 10, Substitution
12. $x = (y \circledast z) \circledast z^{-1}$	2, 7, Substitution
13. $x = y$	11, 12, Substitution
14. $x \circledast z = y \circledast z \rightarrow x = y$	1, 13, CP

If you're so inclined, provide your own proof of T2. There's no solution in the back of the book, so here's a little guidance. First try to show that $(x \circledast \hat{\imath}) \circledast x^{-1} = (\hat{\imath} \circledast x) \circledast x^{-1}$. Then use T1 to show that $x \circledast \hat{\imath} = \hat{\imath} \circledast x$.

Group theory is very basic. It was said above that "group theory may be considered the most fundamental theory in mathematics." But can we find a more basic theory?

Here's one that's more basic: the theory of equality. It has two axioms, $\mathcal{EQ}_1$ and $\mathcal{EQ}_2$.

> $\mathcal{EQ}_1.$ $\forall x(x = x)$
> $\mathcal{EQ}_2.$ $\forall x \forall y(x = y \rightarrow (\varphi xx \rightarrow \varphi xy))$

Axiom $\mathcal{EQ}_2$ conveys the idea that whenever two things are equal (x and y), whatever is true of one is true of the other (whatever is true of x is true of y). (φxy is just like φxx except for having "y" substitute freely in some places where "x" was free, where "free" means not governed by a quantifier.) This is sometimes referred to as Leibniz's law.

The theory of equality has only a few theorems, those we expect from equality. First, we have symmetry: if $a = b$, then $b = a$. Let's call this T1. A proof of this is shown below.

> T1. $\forall x \forall y(x = y \rightarrow y = x)$

1. $a = b$	1, As (1)
2. $\forall x \forall y(x = y \rightarrow (x = x \rightarrow y = x))$	$\mathcal{EQ}_2$
3. $a = b \rightarrow (a = a \rightarrow b = a)$	UI, twice, from 2
4. $a = a$	$\mathcal{EQ}_1$, UI
5. $b = a$	1, (1, 3 MP, 4 MP)
6. $a = b \rightarrow b = a$	1, 4 CP

And we have transitivity: if $a = b$ and $b = c$, then $a = c$. We'll call this T2 and prove it below.

> T2. $\forall x \forall y \forall z \, (x = y \wedge y = z \rightarrow x = z)$

1. $a = b \wedge b = c$	1, As (1)
2. $\forall x \forall y(x = y \rightarrow (x = c \rightarrow y = c))$	$\mathcal{EQ}_2$
3. $b = a \rightarrow (b = c \rightarrow a = c)$	2, Substitution (b/x, a/y)
4. $a = b$	1
5. $b = a$	T1
6. $b = c \rightarrow a = c$	MP
7. $b = c$	1
8. $a = c$	6, 7 MP

Is there any theory weaker than the theory of equality? Sure. Here's one: the theory of quasi-equivalence. Although it seems widely thought that reflexivity follows from symmetry and transitivity, it doesn't. What's the catch? First, let's exhibit the two axioms for quasi-equivalence:

$Q\mathcal{E}_1$. $\forall x \forall y (Fxy \rightarrow Fyx)$
$Q\mathcal{E}_2$. $\forall x \forall y \forall z (Fxy \wedge Fyz \rightarrow Fxz)$

Now, let's examine the quasi-reflexivity theorem: $\forall x (\exists y (Fxy \rightarrow Fxx))$.

Perhaps somewhat surprisingly, we cannot prove reflexivity—that $\forall x (Fxx)$. The catch is that we don't know, for any two elements x and y, whether either Fxy or Fyx holds. This may seem trivially uninteresting, but the actual derivation of quasi-reflexivity might possibly hold some interest. It's surprisingly long: 15 steps, to be exact, in a fussy derivation. Instead of displaying all 15 steps, here are just steps 1, 2, 3, 5, 7, and 15. You should have no trouble filling in the missing steps on your own. Try it for fun, but there's no solution in the back of the book.

To Prove: $\forall x (\exists y (Fxy \rightarrow Fxx))$

1. $\exists y (Fay)$	1, As (1)
2. Fab	
3. $\forall x \forall y (Fxy \rightarrow Fyx)$	$Q\mathcal{E}_1$
...	
5. $Fab \rightarrow Fba$	
...	
7. $\forall x \forall y \forall z (Fxy \wedge Fyz \rightarrow Fxz)$	$Q\mathcal{E}_2$
...	
15. $\forall x (\exists y (Fxy \rightarrow Fxx))$	14, UG

To get genuine reflexivity, add connectivity: "for any two elements, one must bear F to the other."

$Q\mathcal{E}_3$. $\forall x \forall y (Fxy \vee Fyx)$

Chapter 22
Computer Science

Bubble Sort, Comb Sort, and Quicksort

Let's look at some nifty notions in computer science. To sort an array of numbers, often the complicated (recursive) Quicksort is used because it's a fast sort algorithm. In beginning computer science classes, usually Bubble Sort, the slowest of the sorts, is first introduced because it is easiest for students to understand. What Bubble Sort does is compare adjacent elements pair by pair, swapping them when the first of the pair is larger than the second. Here's an example of the largest number "bubbling" up to the end of the list, then the next one, and so on.

Consider the following array of numbers. Since we first want to bubble up to the array's end the largest number, 63, we've made 63 bold so we can follow its progress.

43, **63**, 11, 28, 17, 33

We consider the first two numbers, in positions 1 and 2. In this case, the number at position 2 is larger than the one in position 1, so no swap takes place.

43, **63**, 11, 28, 17, 33

Then we compare the numbers at positions 2 and 3. In our example, 63 is larger than 11, so the two numbers are swapped. The result is that the largest of the first three numbers happens now to be in position 3 (in bold):

43, 11, **63**, 28, 17, 33

We repeat this bubbling up process of comparing adjacent numbers until we reach the end of the array, at which point the largest of all elements is in the end position. Here are the remaining steps that lead up to 63 being at the end:

43, 11, 28, **63**, 17, 33
43, 11, 28, 17, **63**, 33
43, 11, 28, 17, 33, **63**

Of course, we now need to loop through all number pairs to get 43 to bubble up to the next-to-last position. As I'm sure you've guessed, "bubbling up" gives Bubble Sort its name.

In the end, by going through each loop again and again, shortening its length by 1 each time through, we end up with this ascending sorted array:

11, 17, 28, 33, 43, 63

In general, each step in the bubbling-up process is reflected by this partial loop (written in quasi-pseudocode):

For $i = 1$ to $n - 1$ (where $n =$ the number of numbers in the array), compare i with $i + 1$;
if $i + 1 > i$, then swap, diminishing i by 1 each time through the loop.

If you're unfamiliar with computer sorts of any kind, Bubble Sort may seem nifty. It's clear, and its progress can be followed step by step.

But Bubble Sort is slow! It requires us to loop again and again for the number of elements in the list minus 1 until we have but two numbers left. That's roughly n^2 times, where n is the number of elements in the array. The quickest sort used by programmers is Quicksort. It's harder to code and uses recursion, which is not always so easy to implement in so-called "procedural languages." But it happens that a little-known sort called Comb Sort just requires one tiny tweak in the easy-to-code Bubble Sort. That tweak is to compare numbers by a factor of 1.3 instead of comparing adjacent numbers—i.e., those separated by a factor of 1.

Here's how it works. First, let's look at an array of numbers like before. There are 10 numbers in our array. Divide the number 10 by 1.3 and you get what's called a "gap" of 7 by rounding downwards.

Let's try it with the following array:

30, 91, 72, 11, 57, 17, 75, 43, 65, 89

We start at the beginning of the array of numbers, comparing the first with the item one gap further up the set—i.e., the number in the eighth place. The two numbers are as follows:

30, 43

So, don't swap. We continue to the end of the array, then shrink a new gap for the comparisons already made by 1.3, which reduces them. We continue in this way, always going to the end of an array and recalculating the gap, until it's 1, which returns us to the ordinary Bubble Sort.

What's neat about Comb Sort is that it's both an easy sort to code and an extremely fast sort (unless the numbers in the list are already sorted). Recalculating the gap takes almost no time at all. And there are far fewer loops, which consume time. As mentioned, it's almost as fast as the more difficult Quicksort. And it's almost as easy to code as the easiest sort, Bubble Sort. What's strange about Comb Sort is that given its ease and speed, it is virtually never taught in beginning programming classes.

The "Swap" Function in Computer Programming

This presentation of how to "switch" two numbers—or two of anything else—in a computer program is definitely nifty in my opinion, although it's not usually considered important enough to be called a mathematical result. Indeed, it seems that few programmers know of the very simple program Switch. We begin with an ordinary explanation of a related program called Swap.

The fundamental idea can be illustrated by moving furniture in a room. Suppose you wish to move a chair to where a table is, and to move the table where the chair was. It's immediately clear that you cannot just exchange the locations of these two pieces. Supposing you move the chair first, you need some temporary place to put the chair before moving the table to where the chair had been. Then, after that, you move the chair to where the table once was. So, three places are needed in order to swap the chair and table. The same thing is true in computer programming. We use the notation $x \leftarrow y$ to mean that the value of y is being moved to x. It is also possible to perform a mathematical operation on the value of y and send the result to x. For example, $x \leftarrow y + 10$ adds 10 to the value of y and then sends the result to x.

If you wish to swap the values of x and y, you need to do the same thing as you did with furniture. Namely, find a temporary place to first store one of them so the exchange can be made.

Here is a typical Swap program. To check that it works, substitute some numbers for x and y. Let's try 3 and 7, respectively. When Swap is finished, check that the values are now reversed.

```
Swap:
temp ← y
y ← x
x ← temp
```

Now, here's a surprise: unlike when moving furniture, you do not really need to have a temporary location to exchange two values. If you think about using only two variables instead of three, it seems counterintuitive that switching them is possible. That you really can do it seems nifty to me. I hope it seems so to you, too. We'll call this program Switch.

Switch:
$y \leftarrow x + y$
$x \leftarrow y - x$
$y \leftarrow y - x$

Does it really work? Let's mix a positive and a negative number just to check a single example. Make $x = 3$ and $y = -10$.

$x = 3$
$y = -10$
$y \leftarrow 3 + (-10)$ y is now $= -7$
$x \leftarrow y - 3$ x is now $= -10$
$y \leftarrow y - (-10)$ y is now $= 3$

Not an earth-shattering result, but a neat one just the same. One wonders why Switch is never used in teaching computer science and also not used by programmers. It's far more elegant than Swap. I suppose that Swap does its switching in a more obvious way than Switch. Perhaps, besides being nifty, Switch has a somewhat suspicious air to it. But once you see that it works, the suspicion should vanish.

Figure 22.1 Photograph of Alan Turing (1912–1954). Photo credit: The Provost and Scholars of King's College Cambridge 2024.

Turing Machines

One reason for introducing Turing machines is to present a nifty argument that the mathematician and code-breaker Alan Turing produced, which, among other titles, is often called "the unsolvability of the halting problem." There's much I need to explain to make the programs of a Turing machine make sense, even before tackling the halting problem. Fortunately, the details of "programs" that run on a Turing machine are quite nifty unto themselves.

A Turing machine is a simple, abstract computational device intended to help investigate the extent and limitations of what can be computed. We will very roughly explain how a Turing machine works by using an example of a simple program. The example alone does not provide a technical definition of a Turing machine, though we hope it's enough to give the basic idea of how a Turing machine works. The example is to add two numbers. We'll add two plus three in terms of 1s and 0s. We begin with 0011011100, which represents "2 + 3," and we will wind up with 0001111100, which is 5, simply by filling in the blank space between the two numbers. First of all, picture a tape that looks something like a railroad track, as illustrated in Fig. 22.2.

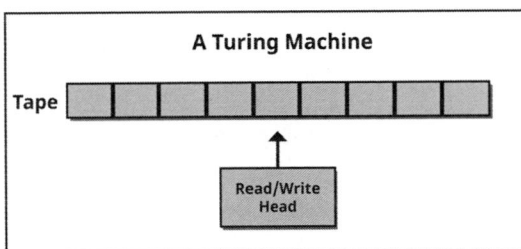

Figure 22.2   Illustration of a Turing machine. Image credit: Ernesto Mora.

We assume that there are various "states" the Turing machine is in, and we represent these states by q with a subscript. A state can move to the right (R) square or the left (L) square and can change 0 to 1 or 1 to 0. The Turing machine's "head" starts at the square with the leftmost 1 and will end similarly—on the leftmost 1. We begin with our Turing machine in state q_0 and proceed as indicated by the six steps below.

1. $<q_0, 1, 0, q_0>$
2. $<q_0, 0, R, q_1>$
3. $<q_1, 1, R, q_1>$
4. $<q_1, 0, 1, q_2>$
5. $<q_2, 1, L, q_2>$
6. $<q_2, 0, R, q_3>$

The first quadruple, $<q_0, 1, 0, q_0>$, tells us to replace the first 1 by a 0 and remain in state q_0. We show this below on our "railroad track." First, we start at the first 1:

0	0	1	1	0	1	1	1	0	0
		↑							

Here is the result of following step 1 (replacing 1 by 0):

0	0	0	1	0	1	1	1	0	0
		↑							

The next quadruple is indicated by $<q_0, 0, R, q_1>$, which means that when a 0 is encountered, move to the right (R) and go to state q_1, and so on.

By following steps 1 through 6, we end up with 5, the sum of $2 + 3$, as indicated below:

0	0	0	1	1	1	1	1	0	0
↑									

Exercise 42: Write a Turing machine program to add 2 to 3 in a more complicated fashion.

Adding two numbers in a more complicated fashion suggests that we could add two numbers several times, which gives us multiplication. Multiplying numbers several times provides exponentiation. In short, it is possible to see that all operations can be accomplished with this simple Turing machine— although, of course, it would take more space and time.

The Unsolvability of the Halting Problem

What Alan Turing proved is that no computer can reliably decide whether a computer operating on a given input will reach a result and halt. We can arrive at this proof by assuming the contrary, that a Turing machine can decide, and we obtain a contradiction from this assumption.

Let H be a Turing machine that decides whether a given Turing machine will halt on a given input. We can express this by $H(T, x)$, where H is for "Halting," T is some given Turing machine, and x is the program that T runs. H, T, and x are all represented by a unique numeric code. H is a Turing machine that evaluates whether T operating on x will halt or whether it will go into an endless loop.

Now we construct a new Turing machine, H', which has H as a subroutine. H' calls H to determine what H does when x is itself H's own program. But H' does just the opposite of what H does. Thus, if H halts, H' goes into a loop; if H loops, then H' halts. In other words, H' calling H to check whether H halts outputs the very opposite that H does.

So, regardless of what H' outputs when it operates on H, it's forced to output the opposite result. So, H cannot exist. Thus, deciding whether a Turing machine halts is impossible.

You might not really care whether a program consisting of a sequence of quadruples matters at all. But Turing proved that his Turing machine could do anything an ordinary digital computer could do.

Turing wasn't the only one to invent methods of computability. Around the same time Turing was writing his programs, several other mathematicians analyzed what it means to be "effectively computable," "recursively computable," or just "mechanically computable"—these terms are all equivalent; each mathematician arrived at a unique formulation of the same basic notion. What's truly interesting is that every single one of these formulations, worked out along extremely different lines than a Turing machine, arrived at provably equivalent notions. Thus, it's commonly supposed that the intuitive notion of what we mean by mechanically computable (and all the other synonyms) has been completely captured by Turing's and these other mathematicians' different constructions.

If you desire something more concrete, consider a personal computer, perhaps the one sitting on your desk. A Turing machine can compute anything your computer can compute—and more, actually, since the tape of a Turing machine is potentially infinite in length and your computer is finitely bounded. You may be baffled by this, since Turing machines are so limited, using just 1s and 0s and moving to the right or left in order to go into another state. But consider this: we have exhibited only a simple adding machine above. But we could write a Turing machine program that multiplies. After all, multiplication is just repeated addition. Writing a "repeat adding" Turing machine is of course more complicated. But once we have multiplication, we can write a Turing machine to do exponentiation. While these programs would be much more complex to write out, the point is not to achieve convenience but generality.

Chapter 23
Undecidability: Gödel and Chaitin

Suppose we could separate truth from provability. This is not to say that something false could be provable; just that something true might not be provable. How could this happen? Well, take some outstanding conjecture about infinite numbers. We know that Euclid proved infinitely many primes, so we know it's true. But suppose we didn't have such a proof. Before he proved it, it may have been thought true yet lacked a proof.

For a concrete example, let's look at one of the astonishing results arrived at by Kurt Gödel (Fig. 23.1)—a 20th century mathematician generally ranked

Figure 23.1 Photograph of Kurt Gödel and Dorothy Morgenstern Thomas at the Institute for Advanced Study. Photo credit: A. G. Wightman from the Shelby White and Leon Levy Archives Center, Institute for Advanced Study (Princeton, New Jersey).

among the most important logicians in history. Gödel concocted a sentence within arithmetic that states that it is unprovable. (Actually, since this sentence is in an axiomatic system, it is more precise to say that the sentence is "not derivable" rather than "not provable.") The polymath John von Neumann wrote: "Kurt Gödel's achievement in modern logic is singular and monumental—indeed it is more than a monument, it is a landmark which will remain visible far in space and time."

To get ultra-technical for a brief moment, Gödel's sentence within a mathematical theory (to be shown incomplete) is

Self(k) $\neg\exists Der(y, \text{Self}(k))$

That is, within the theory itself, Gödel constructed a sentence that we've named (by the number) Self(k), which is true theoretically if and only if there's a derivation y in the theory such that y derives Self(k). Thus, $\neg\exists Der(y, \text{Self}(k))$ says of itself that it's not derivable. Whew!

Let's emerge from the technical nature of the sentence above and just look at it in perhaps a more satisfying way. Gödel wanted to banish any reference to truth from his proof. He did that by showing that truth was undefinable within the theory, but derivability was definable. Calling the above sentence Self(k), we can say that Self(k) is not derivable. Often the letter G for Gödel is used for the sentence Self(k). Sentence G is not derivable, given the theory's consistency, and neither can $\neg G$ be derived, given a somewhat stronger assumption than consistency.

Many people have concluded that Gödel's proof really shows them the truth of $\neg G$, since $\neg G$ does not express the non-derivability of G, which is true (i.e., sentence G is not derivable). Thus, they claim that they can therefore *see* from Gödel's proof that $\neg G$ must be true. But that's not the case. They cannot really *see* the truth of $\neg G$ either, since to see the truth of that sentence requires one to first see the truth of the consistency of the theory, since this is an assumption needed for Gödel's proof. But none of us is in any position to see the truth of the consistency of the theory. In fact, it's theoretically possible that the theory is inconsistent. Gödel showed only that after assuming it is consistent, there's a sentence that can neither be proved nor disproved. One way to state his result is that given a sufficiently strong system of arithmetic, there are more truths than proofs.

To head off one misinterpretation of Gödel's proof of the incompleteness or undecidability of arithmetic, we can state that it definitely does not hold for any axiomatized theory. As in the sentence above, it holds for "a sufficiently strong system of arithmetic"—or sufficiently weak, if you like. Recall Q from Chapter 5. Q is weaker than Peano arithmetic. Q is so weak that $0 + x = x$ cannot be proved in it (although $x + 0 = x$ can be proved). Yet Gödel's proof holds in Q.

Figure 23.2 Photograph of Raymond Smullyan playing the violin to his dog, probably around 1929–1930. Likely photographer was his father. Photo credit: The Raymond Smullyan Society.

Smullyan, Again

Another diabolically clever enigma by Raymond Smullyan is a clear counterpart to Gödel's theorem and may help to elucidate it.

The symbol "☆" names a sentence:

☆   ☆ is unprintable.

First, we need to define some things. An expression is defined as any finite, non-empty string of the following five symbols:

¬   *P*   *N*   (   )

These symbols are printable by a computer that can sooner or later print everything it's capable of printing. A norm of one of the five expressions is that X is $X(X)$. In other words, the norm of $P\neg$ is $P\neg(P\neg)$. For any given expression X, a sentence is defined as one of the following four forms:

1. $P(X)$
2. $PN(X)$
3. $\neg P(X)$
4. $\neg PN(X)$

P stands for "printable," N stands for "the norm of," and $\neg$ stands for "not." $P(X)$ is defined as True if and only if X is printable. $PN(X)$ is True if and only if the norm of X is printable. $\neg P(X)$ is True if and only if X is not printable. And $\neg PN(X)$ means that the norm of X is not printable.

Smullyan mentions that this small system is capable of a kind of self-reference. Paraphrasing him, the machine prints out various sentences about what it can and cannot print, and therefore it describes its own behavior.

Now, we are told that the machine will eventually print only True sentences— that is, it will print no False sentences. The tantalizing question Smullyan puts before us is whether the machine can print *all* True sentences.

Before continuing with this question, notice the comparison between "print" and "prove." In various logical systems, one can prove only True sentences. The question then arises for such a given system whether it is possible to prove all True sentences. The analogous question for Smullyan's print system is whether it can *print* all True sentences (we know it can print only True ones).

The answer is: No!

Exercise 43: Find a True, Gödel-type sentence that the machine cannot print. Smullyan hints that the answer is to find a sentence that asserts its own non-printability just like Gödel's G.

Another useful way to think of Gödel's G is as a counterpart of the liar paradox, which we examined in Chapter 13. Both express the undecidability of being proven either True or False (given the consistency of the systems in which they have been formulated).

Chaitin's Theorem

Chaitin's theorem refers to a result in an exciting, relatively new field called algorithmic information theory, which Gregory Chaitin co-founded together with Russian mathematician Andrey Kolmogorov.

Chaitin begins with the notion of the complexity of a string of 0s and 1s. For example, the string 01010101010101 is not very great, since we can see that the string is just 01 written seven times. The string consisting of 10,000 representations of 01 is not much more complex, as it can be generated by the command "print '01' 10,000 times." On the other hand, the string 1101011001011100100001 seems much more complex, as the simplest way to generate it seems to be the command "print '1101011001011100100001.'"

To be slightly more precise, the command to generate the first string required only 18 symbols (including spaces). To generate the second took 23, and the third required 36 symbols. The language in which commands like this are written could be a computer language, but for simplicity we will stick with ordinary English.

Even though Chaitin's theorem is a mastery of technical detail, we can already appreciate some of his ideas in an intuitive way. Suppose that in our command language we permit expressions like "print the first string of complexity > 41." Since the complexity of the command itself is 41—i.e., there are 41 characters in the command (excluding the final period)—we are saying to print the first string that *cannot* be generated by this very command.

Again, note the similarity to the liar paradox. We will call this "Chaitin's paradox," which is to command that something be done that cannot be consistently done. Suppose that such a string *were* generated. Then its complexity would be 41 or less since a command of length 41 can generate it. On the other hand, if the command is carried out faithfully, then the string's complexity must be *greater* than 41 since the command *states* that the string must have complexity greater than 41.

Formally, Chaitin's paradox can be carried out by coding the symbols into binary strings, and then by defining a derivative checker (d-c) to print a number if it is a code number of a derivation in N (the theory of natural numbers). We are supposed to know that the process of coding and decoding sentences of N is mechanical, and so is the process of checking whether a string is a derivation. Further, this derivation checker (d-c) has a unique length that can be measured in terms of the number of symbols the procedure employs. (This can be coded so that the number of each procedure is unique.)

Suppose that the length (somehow measured) of the d-c is l. Then define the complexity of a coded theorem number to be the length of the shortest procedure that prints that number. Since there are infinitely many theorems, there must be theorem numbers whose complexity is greater than l. In fact, there are infinitely many true sentences of the following form: "the complexity of k is greater than l." Let j be the number of such a sentence, where the complexity of j itself is greater than l. (Since there are infinitely many true sentences of that form, there must be infinitely many whose code numbers are greater than l.) So, suppose that d-c checks j. Since d-c does not make errors in checking theorems (by strong soundness), it will correctly arrive at the fact that the complexity of k is greater than l. Or will it?

For d-c to carry out its task, it must print j, which can only be printed by procedures of greater length than d-c itself (since the complexity of j is greater than l). Thus, it cannot consistently print j, and thus it cannot determine whether or not the complexity of k is in fact greater than that of l. Since there are infinitely many of such true sentences, there are infinitely many whose code numbers are greater than l that d-c cannot analyze.

Thus, Chaitin's paradox can be produced in N, showing, as Gödel did, that N is incomplete. Chaitin's paradox, however, gives us different information than Gödel's theorem. It tells us that every procedure has a given complexity that it cannot go beyond. This fuels speculation about the limitations of our own capacity to understand the world in which we live. Suppose that our degree of complexity is h (for humans). Chaitin's theorem seems to tell us that we cannot correctly analyze the complexity of anything beyond h. This should *not* be confused, however, with our being able to understand that some fact of complexity greater than h is true. We may well be able to understand something much more complex than ourselves. There are infinitely many theorems of N that can be derived in N, and thus there are infinitely many of them beyond any particular complexity. What we cannot determine in N is that the complexity of a theorem is beyond the complexity of the procedure used to check its complexity. And thus, Chaitin's theorem seems to tell us that we are in no position to evaluate the complexity of anything greater than our own minds.

Solutions to Exercises

Solution to Exercise 1: Using Gauss' formula, $S_n = n[2a + (n - 1)d]/2$, sum a series of eight numbers that begins with 3, where each subsequent number is 7 greater than its precedent.

The number series is 3, 10, 17, 24, 31, 38, 45, 52.
$n = 8$, $a = 3$, and $d = 7$.
The sum is 220.

Solution to Exercise 2: Just for fun, suppose that Gauss was asked a slightly different question: "How many numbers in a series would you need to reach 565 (S_n), where the first number a is 7, then continuing to add some large d, say 53?"

Begin with the formula $S_n = n[2a + (n - 1)d]/2$.
$S_n = 565$, $a = 7$, $d = 53$.
$565 = n[14 + (n - 1)53]/2 = n[53n - 39]/2$.
$1130 = n[53n - 39]$
$53n^2 - 39n - 1130 = 0$.
$n = 5$ or $n = -226/53$

Solution to Exercise 3: The problem is to provide a proof by mathematical induction that $1 + 2 + 3 + ... + n = [n(n + 1)]/2$. I'll start the proof, issue a warning, then leave the rest of it for you to work out.

PROOF:
Base Case: $1 = (1(1 + 1))/2$, since $(1(1 + 1))/2 = 1(2)/2$.
IH: $1 + 2 + 3 + ... + k = (k(k + 1))/2$.
To Prove: It's true for $k + 1$ substituted in the equation above. That is, prove that $1 + 2 + 3 + ... + k + (k + 1) = ([k + 1]([k + 1] + 1))/2$.
Proof: ...

By IH, you already know that $1 + 2 + 3 + \ldots + k = (k(k + 1))/2$. So, substitute the right-hand side of this equation in the one above. That yields $[k(k + 1)]/2 + (k + 1)$. Now prove that it equals $([k + 1]([k + 1] + 1))/2$. In other words, prove that $[k(k + 1)]/2 + (k + 1) = ([k + 1]([k + 1] + 1))/2$. You will need to manipulate these expressions using some elementary algebra.

Here's the warning that was promised above: in proving $[k(k + 1)]/2 + (k + 1) = ([k + 1]([k + 1] + 1))/2$, you must manipulate each side independently. That is, the left-hand side (LHS) is $[k(k + 1)]/2 + (k + 1)$, and the right-hand side (RHS) is $([k + 1]([k + 1] + 1))/2$. If you transfer any element between the LHS and the RHS anywhere in your proof, you're assuming that they're equal, which is what you wish to prove, making your proof circular. When you get the LHS to equal the RHS by algebraic maneuvering, you will have finished the proof. This may seem difficult, but it's not. Just as long as you remember a little algebra.

Solution to Exercise 4: Prove for $n \geq 4$: $3^n > 2n^2 + 3n$.

Here's an outline of a solution. First, assure yourself that the inequality fails for 1, 2, and 3. Then, plug in 4 and attest that the inequality is true. Assume by IH that $3^k > 2k^2 + 3k$ holds.
Now, multiply both sides by 3, yielding $3(3^k) > 3(2k^2 + 3k)$, which is $6k^2 + 9k$.
Then by some manipulations, we get $2(k + 1)^2 + 3(k + 1)$, and we're done.

Solution to Exercise 5: Prove that 21 divides $4^{n+1} + 5^{2n-1}$ for $n \geq 1$. The base case has already been shown. We now need to prove that $4^{(k+1)+1} + 5^{2(k+1)-1}$ is divisible by 21, i.e., that $4^{k+2} + 5^{2k+1}$ is divisible by 21. By IH, we have that $4^{k+1} + 5^{2k-1}$ is divisible by 21.

$$4^{(k+1)+1} + 5^{2(k+1)-1}$$
$$= 4^{k+2} + 5^{2k+1}$$
$$= 4 \cdot 4^{k+1} + 5^2 \cdot 5^{2k-1}$$
$$= 4 \cdot 4^{k+1} + 25 \cdot 5^{2k-1}$$
$$= 4 \cdot 4^{k+1} + (4 + 21) \cdot 5^{2k-1}$$
$$= 4 \cdot 4^{k+1} + 4 \cdot 5^{2k-1} + 21 \cdot 5^{2k-1}$$
$$= 4(4^{k+1} + 5^{2k-1}) + 21 \cdot 5^{2k-1}$$

We know by IH that $4^{k+1} + 5^{2k-1}$ is divisible by 21, and so is $21 \cdot 5^{2k-1}$.

Solution to Exercise 6: Prove by mathematical induction that a set of n elements has 2^n subsets for $n \geq 0$.

PROOF:
Base Case: Take a set with 0 elements. $2^0 = 1$. This set as a 1 element subset, namely $\emptyset$, the empty set. Thus, the theorem is true when $n = 0$.

IH: The induction hypothesis is that any set with k elements has $2k$ subsets. We need to prove that a set of $k + 1$ elements has $2k + 1$ subsets. Let subset A have $k + 1$ elements: a_1, a_2, a_3, ..., a_{k+1}. And let $A' = A - \{a_{k+1}\}$. So, subset A' has only k elements: a_1, a_2, a_3, ..., a_k. Thus, the induction hypothesis holds for A', since it has 2^k subsets.

Take subset B such that $B =$ the union of B' and $\{a_{k+1}\}$, where B' is a subset of A'. By the induction hypothesis, there are also 2^k such subsets B'. Since there are 2^k subsets of A' and 2^k subsets of B', the total number of subsets of A equals $2^k + 2^k = 2(2^k) = 2^{(k+1)}$. This proves by induction that any set of n elements has 2^n subsets.

Solution to Exercise 7: Prove by mathematical induction that $1^3 + 2^3 + 3^3 + \ldots + n^3 = n^2(n + 1)^2/4$.

PROOF:
Base Case: $n = 1$. $1^3 = 1^2 \cdot [1^2 \cdot (1 + 1)^2]/4 = (1^2)(2^2/4) = 4/4 = 1^3$.
IH: $1^3 + 2^3 + 3^3 + \ldots + k^3 = [k^2(k + 1)^2]/4$.
To Prove: $1^3 + 2^3 + 3^3 + \ldots + k^3 + (k + 1)^3 = [(k + 1)^2 \cdot (k + 2)^2]/4$.
Proof: $[k^2(k + 1)^2]/4 + (k + 1)^3 = [(k + 1)^2 \cdot (k + 2)^2]/4$.
Manipulating the left-hand side of the equation: $[k^2(k + 1)^2]/4 + 4(k + 1)^3/4 = [k^2(k + 1)^2] + 4(k + 1)^3]/4$
$= (k + 1)^2 \cdot [k^2 + 4(k + 1)]/4$
$= [(k + 1)^2 \cdot (k + 2)^2]/4$.

Solution to Exercise 8: Prove for $n \geq 4$ that $n! > 2^n$.

PROOF:
Base Case: $n = 4$. $4 \cdot 3 \cdot 2 \cdot 1 = 24$. $2^4 = 16$. $24 > 16$.
IH: If $k \geq 4$, $k! > 2^k$
To Prove: If $(k + 1) \geq 4$, $(k + 1)! > 2^{k+1}$
Proof: $2^{k+1} = 2 \cdot 2^k = 2^k + 2^k$. And $(k + 1)! = (k + 1)k! = k \cdot k! + k!$
We know by IH that $k! > 2^k$.
$k \cdot k! + k! > k \cdot 2^k + 2^k$. And we know that $k \geq 4$, so we can add this comparative statement…
$k \cdot k! + k! > k \cdot 2^k + 2^k \geq 2^k + 2^k$. By taking the strongest inequality, this can be reassembled into…
$(k + 1)! > 2^{k+1}$.

Solution to Exercise 9: Prove that $7^n - 1$ is divisible by 6 for each positive integer n.

PROOF:
Base Case: Prove for $n = 1$. $7^1 - 1 = 7 - 1 = 6$.
Assume by IH that $7^k - 1 = 6 \cdot m$ for some positive integer m.
To Prove: $7^{k+1} - 1$ is divisible by 6.

$$7^{k+1} - 1 = (7 \cdot 7^k) - 1$$
$$= (7 \cdot 7^k) - 1 - 7 + 7 = 7(7^k - 1) + 7 - 1 = 7(7^k - 1) + 6$$
$$= \text{(by IH) } 7(6m) + 6 = 6(7m) + 6(1) = 6(7m + 1).$$

Solution to Exercise 10: Prove that $7n + 5$ is divisible by 6 for each positive integer n.

One easy way to do this is just to note that $7^n + 5 = (7^n - 1) + 6$, and that $(7^n - 1)$ has already been proved to be divisible by 6. Otherwise, proceed by induction.

PROOF:
Base Case: Prove for $n = 1$. $7^1 + 5 = 7 + 5 = 12 = 6(2)$.
Assume by IH that $7^k + 5 = 6m$ for some positive integer m. Then,
Prove: $7^{k+1} + 5$ is also divisible by 6.
$$7^{k+1} + 5 = (7 \cdot 7^k) + 35 - 35 + 5 = 7(7^k + 5) - 35 + 5 = 7(7^k + 5) - 30$$
$$= \text{(by IH) } 7 \cdot 6m - 30 = 7 \cdot 6m - (6 \cdot 5) = 6(7m) - 6(5) = 6(7m - 5).$$

Solution to Exercise 11: Prove by induction for $x \geq 2$, $n \geq 1$, that $x^n - 1$ is divisible by $x - 1$.

PROOF:
Base Case: Prove for $n = 1$ and $x = 2$. Show that $2^1 - 1$ is divisible by 1.
$2^1 - 1 = 1$.
Assume by IH that $x^k - 1$ is divisible by $x - 1$. Let $(x^k - 1)$ be $(x - 1)m$, where m is some positive integer.
To Prove: $x^{k+1} - 1$ is divisible by $x - 1$.
Proof: $x^{k+1} - 1 = x \cdot x^k - 1 = x \cdot x^k - 1 - x + x = x(x^k - 1) - 1 + x$
$= \text{(by IH) } x(x - 1)m - 1 + x = xm(x - 1) + (x - 1) = m(x + 1)$.

A simpler way to prove that $x^n - 1$ is divisible by $x - 1$ would be to appeal to the factor theorem $x^n - 1 = (x - 1)(x^{n-1} + x^{n-2} + \ldots + x^2 + x + 1)$, which is a generalization of these examples:

$$x^2 - 1 = (x - 1)(x + 1)$$
$$x^3 - 1 = (x - 1)(x^2 + x + 1)$$
$$x^4 - 1 = (x - 1)(x^3 + x^2 + x + 1)$$
$$x^5 - 1 = (x - 1)(x^4 + x^3 + x^2 + x + 1)$$
$$x^6 - 1 = (x - 1)(x^5 + x^4 + x^3 + x^2 + x + 1)$$

However, the factor theorem isn't really necessary. For the theorem we're interested in, we need to prove only that $x^n - 1 = (x - 1)m$, where m is some integer ≥ 1. But we've already proved this above.

Solution to Exercise 12: Prove $2^n > n^2$ for $n > 4$. If you get stuck, prove this first: (a): for $n > 2$, $2^n > 2n + 1$. Then, use (a) to help prove (b): $2^n > n^2$ for $n > 4$.

(a) For $n > 2$, prove $2^n > 2n + 1$.
   Base Case: $n = 3$. $2^3 > 2(3) + 1$. $8 > 7$.
   IH: $2^k > 2k + 1$.
   To Prove: $2^{k+1} > 2(k + 1) + 1$.
   Proof: $2 \cdot 2^k > 2k + 3$.
   By IH, $2 \cdot 2^k > 2(2k + 1)$.
   So, $2 \cdot 2^k > 4k + 2$ and (for $k \geq 1$) $4k + 2 > 2k + 3$.
   Thus, $2 \cdot 2^k > 2k + 3$.

(b) For $n > 4$, prove $2^n > n^2$.
   Base Case: $n = 5$. $2^5 = 32$. $5^2 = 25$. $32 > 25$.
   IH: $2^k > k^2$.
   To Prove: $2^{k+1} > (k + 1)^2$.
   Proof: $2 \cdot 2^k > k^2 + 2k + 1$.
   $2^k + 2^k > k^2 + 2k + 1$.
   By (a), $2^k > 2k + 1$.
   And by IH, $2^k > k^2$.
   Thus, $2^k + 2^k > k^2 + 2k + 1$.

Solution to Exercise 13: Check that each one of $(2 + 1)$, $(2 \cdot 3 + 1)$, and $(2 \cdot 3 \cdot 5 + 1)$ is itself prime. Does continued multiplication of these primes plus 1 always reach a new prime? If not, where exactly does the first non-prime—i.e., the first composite—show up in $(p_1 \cdot p_2 \cdot \ldots \cdot p_n) + 1$, starting with 2 as p_1?

> The first composite is 30,031, which is arrived at by multiplying $2 \cdot 3 \cdot 5 \cdot 7 \cdot 11 \cdot 13 + 1$. 30,031 is also the product of 59 and 509. 59 is the first new prime (509 just happens to be prime as well, but it needn't be.) As it happens, the next two numbers arrived at in this way are also composite. $2 \cdot 3 \cdot 5 \cdot 7 \cdot 11 \cdot 13 \cdot 17 + 1 = 510{,}511 = 19 \cdot 97 \cdot 277$. And $2 \cdot 3 \cdot 5 \cdot 7 \cdot 11 \cdot 13 \cdot 17 \cdot 19 + 1 = 9{,}699{,}591 = 347 \cdot 27{,}953$.

Solution to Exercise 14: Give an alternative proof to Euclid's proof of the infinity of primes by assuming that the largest prime is p and using $N = p! + 1$.

> $p!$, or p factorial, is divisible by all primes less than or equal to p. Thus, $p! + 1$ is not divisible by any of these primes. Hence, N is either a prime greater than p or a composite divisible by a prime greater than p. In each case, there's another prime.

Solution to Exercise 15: Here's a complex puzzle of Lewis Carroll's:

   1. None of the unnoticed things met with at sea are mermaids.
   2. Things entered in the log as met with at sea are sure to be worth remembering.
   3. I have never met with anything worth remembering when on a voyage.

4. Things met with at sea that are noticed are sure to be recorded in the log.

∴ ?????

One answer to the puzzle is: "I have never met with a mermaid at sea." Now that you know this, see whether you can validly deduce it yourself.

Solution to Exercise 16: If the last exercise wasn't too hard, try this one:

1. The only animals in this house are cats.
2. Every animal is suitable for a pet that loves to gaze at the moon.
3. When I detest an animal, I avoid it.
4. No animals are carnivorous unless they prowl at night.
5. No cat fails to kill mice.
6. No animals ever take to me except those in this house.
7. Kangaroos are not suitable for pets.
8. None but carnivora kill mice.
9. I detest animals that do not take to me.
10. Animals that prowl at night always love to gaze at the moon.

∴ ?????

The answer is "Kangaroos are avoided by me." An easy way to see it is to recognize that all of the other terms appear twice and, in so doing, cancel each other out. Only "kangaroos" and "avoided by me" appear once.

Solution to Exercise 17: Why is the following syllogism valid, or why is it not?

1. Everyone loves a lover.
2. Romeo loves Juliet.

∴ Iago loves Othello.

This syllogism by the amazing Raymond Smullyan is not too different from his earlier one that arrives at the surprising conclusion, "I am my own baby." Except this one is weirder and more unnatural, for neither Iago nor Othello even appears in the premises.

This time we follow Smullyan word for word, supplementing his explanation by italicized commentary. "Since Romeo loves Juliet (second premise), Romeo is a lover." *In this argument, we define a "lover" as anyone who loves anyone.* "Since Romeo is a lover, everyone loves Romeo (by the first premise). Since everyone loves Romeo, everyone is a lover." *Recall the definition that anyone who loves anyone is a "lover."* "Since each person is a lover, everyone loves that person (by the first premise). Thus, it follows that everyone loves everyone! In particular, Iago loves Othello."

Note once again that in the above argument, everyone who loves anyone is a lover. Consider Hitler. Hitler was married briefly to Eva Braun and presumably loved her during that period of time. But I think we'd shrink from calling Hitler a lover just because he (may have) loved Eva Braun. Examine the argument with "Hitler loves Eva" replacing the second premise and see whether you think the argument is still valid.

Solution to Exercise 18: Modify the square produced by Socrates' pupil such that you get a square whose area is 8, half the area of the one arrived at by his method.

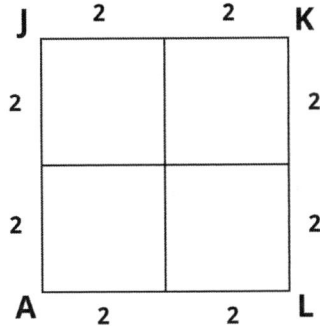

Figure 24.1 Squares: Answer to Plato's problem (Meno's slave). Image credit: Ernesto Mora.

The solution is to draw diagonals on square AJKL, shown below. The area of square AJKL is 16. The area of the square formed by connecting the diagonals is 1/2 the area of AJKL, making its area 8. And we're done!

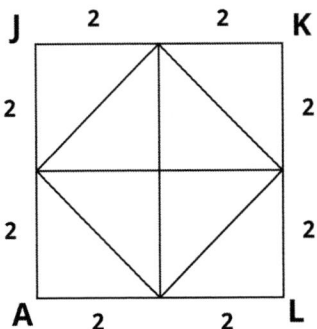

Figure 24.2 Diagonals: Answer to Plato's problem (Meno's slave). Image credit: Ernesto Mora.

Solution to Exercise 19: Convert 0.142857142857... to a fraction in lowest terms.

0.142857142857 converted to a fraction in lowest terms is 1/7.

Solution to Exercise 20: Suppose you have three cards. One of them is red on both sides, one is green on both sides, and the third is red on one side and green on the other. If you grab a card randomly and one side is red, what are the chances that the other side is red as well?

Somewhat surprisingly, the answer is not 1/2; it's 2/3. One way to see this is to note that there are three ways to pick a red card: red-red (that's two distinct ways, because you can pick either one of the two sides that are red) and red-green, which is the third way. So, when you've picked a red card, you may have picked the red side of the red-green card, or you may have picked one of the two sides of the red-red card. That gives you two possibilities out of three. (The third possibility is the red-green card.) Therefore, the probability that the other side is red is 2/3.

Solution to Exercise 21: In a town lives a barber who shaves all men who do not shave themselves. Who shaves the barber?

First, suppose that the barber is a woman. Nothing in the statement of the problem rules that out. According to the problem statement, she shaves all the men in town who don't shave themselves. But the question is: who shaves her? If we wish, we can suppose that she shaves herself—her legs, perhaps. But does she have to shave herself? No. She shaves all men who do not shave themselves, but there's nothing to demand that she, a woman, shaves herself. She can do whatever she wants.

But suppose that the barber in town is a man. Then he must shave every man who does not shave himself. If he does not shave himself, then he does, and if he does... what? He just does; he must shave himself according to the problem description. He cannot grow a beard. But does this raise a problem? No, he just shaves and that's the end of it. If he didn't shave himself then, by statement of the condition of the problem, he must shave himself. Thus, he must shave himself, period. But suppose the condition were that the barber shaves all *and only* those men who do not shave themselves. Then he shaves himself only if he does not (and if he does not, then he does). Now, this is a genuine contradiction. There cannot be such a barber on pain of contradiction, which is the usual analysis of the paradox of the barber.

Perhaps it should be mentioned that the apparent conundrum presented by "there is a barber in the town who shaves all men who do not shave themselves" is often cited as paradoxical, which it's not, as the above analysis reveals.

Solution to Exercise 22: 5/12 of voters prefer A to B, and 7/12 prefer B to C. How many must prefer A to C (assuming transitivity)?

The answer is none! It could be that all five voters who prefer A to B prefer C to A. And it could be that all seven voters who prefer B to C prefer them both to A.

Solution to Exercise 23: Find the two sets of cubes that each sum to 1729.

The two sets of cubes are {10 and 9} and {12 and 1}.
$10^3 \, (= 1000) + 9^3 \, (= 729) = 1729.$
$12^3 \, (= 1728) + 1^3 \, (= 1) = 1729.$

Solution to Exercise 24: How many rabbit pairs appear at the next level? Then the one after that? What is the generalized formula for finding the number of rabbits at any level?

Thirteen rabbit pairs appear at the seventh level. There are 21 at the eighth level. Have you seen the pattern yet? The sum of rabbit pairs in any month n can be found by adding the pairs of rabbits alive in the previous month and the number of new baby rabbit pairs.

To generalize this into a formula, we can call the number of rabbit pairs at the end of the nth month F_n.

The number of new rabbit pairs in month n must be equal to the number of pairs of rabbits who are mature enough to give birth. This is the number of rabbit pairs alive two months previous: F_{n-2}.

Now we need to add the number of new rabbits to the number of surviving rabbits. How do we know the number of surviving rabbits in month n? Since no rabbits die, it's the same as the number from the month before. So, F_{n-1}.

Thus, we have the following formula: $F_n = F_{n-1} + F_{n-2}$.

Solution to Exercise 25: In pseudocode, write a short program in which the Fibonacci numbers are listed from 1 to 100.

Programming the Fibonacci series isn't hard. Write pseudocode for a program containing a recursive function that prints the Fibonacci numbers from 1 to 100.

```
FUNCTION Fib(n)
If n = 1 or n = 2, then
A ← 1.
Else A ← Fib(n – 1) + Fib(n – 2).
For i ← 1 to 100,
Print A.
```

Solution to Exercise 26: A man drives his car 30 miles per hour for 60 miles in going from point A to point B. How fast must he travel on his return journey to average 60 miles per hour for the total distance?

It's impossible! For the man to travel 60 miles from A to B at 30 mph takes 2 hours. If it's already taken him 2 hours to get from A to B, that leaves no time for his return trip. The most common incorrect answer is 90 mph (because half of $90 + 30 = 60$).

Solution to Exercise 27: Two trains 100 miles apart are chugging toward each other on the same track. Train A is traveling toward Train B at 20 miles per hour and Train B is traveling toward train A at 20 miles per hour. A fly flies back and forth between the trains at 40 miles per hour, going from A to B, then B to A, and so forth until the fly gets crushed when the trains collide. How many miles does the fly fly?

The two trains have a combined speed of 40 miles per hour, which means it will take them 2.5 hours to cover the 100 miles between them. From here, the answer is obvious. The fly travels 40 miles per hour for 2.5 hours: 40 miles per hour • 2.5 hours = 100 miles.

Solution to Exercise 28: What is the next number in the following sequence: 1, 2, 4, 8, 16, __?

Before we try to provide an answer for the sequence above, let's look at another problem and a method for its solution.

Consider the sequence 1, 6, 13, 24, 39, 58, __. What is the next number? Is there a method we can use? Yes, but this method doesn't lead to deductive certainty. Remember, any number can follow any series of numbers. Suppose we decide to follow the given numbers by the same numbers in reverse order. So, 1, 6, 13, 24, 39, 58 would be followed by 58, 39, 23, 13, 6, 1 (and then its reverse, which is the same series we started with). There is no absolute proof that we are not "continuing in the same way."

Actually, as you've probably noticed, adding 5 to 1 gives us 6 for the second number. Then add 2 to 5 and get 7; add 7 and get 13 as the third number. Then add 9 (which is $7 + 2$) to the third, getting 24, etc.

Now, we tackle the problem of finding the likeliest number in the sequence: 1, 2, 4, 8, 16, __. Surprisingly, the best answer is 31, not 32.

One way to show this is to use the method of repeated subtraction. This method is applied by listing the series of numbers in a line, then subtracting the smaller number from the larger number to its right, then placing the difference between them on the line below. For instance, in the top row subtract 1 from 2, then place 1 between them on the row below. Moving to the right, subtract 2 from 4, placing 2 below and between them. Repeat these subtractions until all numbers on a single line are equal. Then, we can see that the next number after 16 is 31, and the next after that is 57. Neat, isn't it? However, this method has its limitations and qualifications.

1		2		4		8		16		31		57
	1		2		4		8		15		26	
		1		2		4		7		11		
			1		2		3		4			
				1		1		1				

There is a formula for arriving at 31. But it's much more complex than using powers of 2. The formula gives the value for the number of regions on a circle cut off by chords that do not intersect in more than two points. Here's how it starts. If you wish to follow along, draw a large circle— actually, you had better draw a huge one! Now, put a single point on the circumference of this circle. One point on the circumference doesn't break up the circle at all. So, there's just a single region, the entire circle. Now, draw another point on the circumference and connect it with the first one. Two points break the circle into two regions. With a third point connected to your previous two, there are four regions.

So far, so good. Four connected points yields eight regions, and five gets you 16. But after getting 16 regions for five points, there's a deviation. Instead of 32 regions for six points, you'll only get 31. If the circle you drew was large enough and you chose your points carefully, you'll be able to see this. If not, check the circle in Fig. 24.3 with six interconnected points breaking up the circle into only 31 regions. See whether you can locate the 31st region in the circle below. (Hint: it's the tiniest one.)

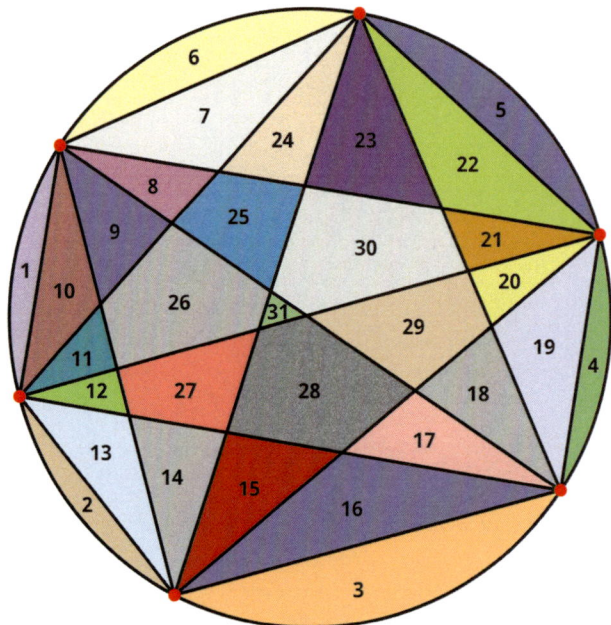

Figure 24.3 Six points on a circle. Image credit: Ernesto Mora.

For any number of points n, the formula for the number of regions is
$r = (n^4 - 6n^3 + 23n^2 - 18n + 24)/24$.

A truncated version of Pascal's triangle will also give this result. Look at the diagram below. At the top, there's a 1. Below that is $1 + 1 = 2$. Then $1 + 2 + 1 = 4$. Then $1 + 3 + 3 + 1 = 8$. Then $1 + 4 + 6 + 4 + 1 = 16$. Then $1 + 5 + 10 + 10 + 5 = 31$. Then $1 + 6 + 15 + 20 + 15 = 57$. Then $1 + 7 + 21 + 35 + 35 = 99$. And so on...

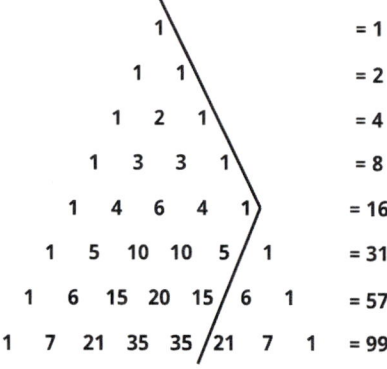

Figure 24.4 Pascal's triangle: split. Image credit: Ernesto Mora.

Solution to Exercise 29: See whether you can extend this interesting sequence: 1, 2, 2, 1, 1, 2, 1, 2, 2, 1, 2, ...

The sequence above is called the Kolakoski sequence and proceeds by defining itself. Here's how: first there's a 1, which appears once (due to the 1 itself) without repeating. Then 2. Since the 2 appears next, the 2 should be repeated twice. Then, there's a 1, which, according to the 2 preceding it should also be repeated twice. The 1 indicates that the next number, a 2, is just repeated once. The idea is that there is a sequence of alternating 1s and 2s. Each of these terms describes how many times, once or twice, a 1 or 2 should be repeated. Here's a longer segment of the Kolakoski sequence: 1, 2, 2, 1, 1, 2, 1, 2, 2, 1, 2, 2, 1, ...

Solution to Exercise 30: Solve the "two bottles" problem by juggling the two bottles or by solving the equation and then applying that solution to the problem itself.

One way to solve this problem is to simply experiment with exchanging liquid in the two bottles until you have the desired 6 quarts in one of them. However, working backwards may simplify the problem. Notice: there's a good chance that the 7-quart bottle will have the 6 quarts in the end. To achieve this, the 11-quart bottle should have 10 quarts of water in it and the 7-quart bottle should be full. Thus, when you pour out the 7-quart bottle into the 11-quart bottle, the 11-quart bottle will be full when the 7-quart bottle has 6 quarts in it. Backing up one more step requires that the 11-quart bottle contain 3 quarts, so that when you pour 7 quarts into it, the 11-quart bottle will have 10 quarts, which is what we wanted for the second step from the end. Skipping over what comes next when we work backwards, we can go instead to the first step and begin pouring. At the first step (bypassing the state when both bottles are empty), you should fill the 7-quart bottle and then (for the second step) pour all of it into the 11-quart bottle, leaving no quarts in the 7-quart bottle and 7 quarts in the 11-quart bottle.

Note that in the previous paragraph, we said that "there's a good chance that the 7-quart bottle will have the 6 quarts in the end." But what happens if we proceed in the same way, from bottom to top, supposing that instead there are 6 quarts of water in the 11-quart bottle? I'm not going to investigate this possibility because, if it works, I know that it will be more complex than the way we've approached it. However, you can proceed in this way and see what happens.

Another approach to the problem would be to solve the Diophantine equation $7x - 11y = 6$. Although we learn from elementary algebra that it's impossible to solve a single equation with two unknowns, we now know that's not actually true. Our algebra teachers misled us. For many Diophantine equations, there is a method to solve them, and this holds for linear Diophantine equations. A linear equation is an equation with no exponent greater than 1.

Returning to the bottle problem, one way to find a solution by not actually solving the equation $7x - 11y = 6$ would be to substitute small numbers for x, then see what positive whole numbers y must be. The first solution you'll arrive at is $x = 4$ and $y = 2$. ($4 \cdot 7 = 28$, which is 6 more than $2 \cdot 11$, which is 22.) So, filling up the 7-quart bottle four times and the 11-quart bottle two times will tell you something, but not everything you want to know about how to end up with 6 quarts in one bottle after pouring water from bottle to bottle. What we really want to know is how we should pour the water between the bottles, and providing values for x and y doesn't really help us much in knowing that.

So, we return to the method presented earlier, starting from the bottom value of 6 quarts in the 7-quart bottle and working up, while also working from the top and working down, just hoping that the two will meet. This seems almost as laborious as simply experimenting by pouring water from bottle to bottle. Let me just provide the simplest answer. There are possibly others, since Diophantine equations are not necessarily uniquely solvable—some could have many answers, and some might not have any. Without further ado, here's a chart for the simplest solution:

7-quart bottle	11-quart bottle
7	0
0	7
7	7
3	11
3	0
0	3
7	3
0	10
7	10
6	11

Note that the 7-quart bottle is filled four times, and the 11-quart bottle is filled only twice, which fits the equation where $x = 4$ and $y = 2$.

Solution to Exercise 31: Prove the midpoint theorem: in any triangle, the line joining the midpoint of two sides is parallel to the third side and equal to one half of it.

Given: triangle ABC, where $AD = BD$ and $AE = CE$.
To Prove: DE is parallel to BC and $DE = 1/2$ BC.

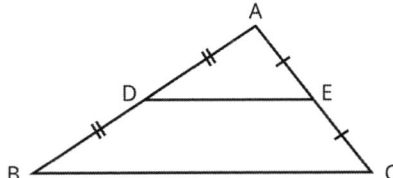

Figure 24.5 Triangle ABC with DE midpoints. Image credit: Ernesto Mora.

Construct a line parallel to AB and let F be the intersection of that line with the extension of DE.

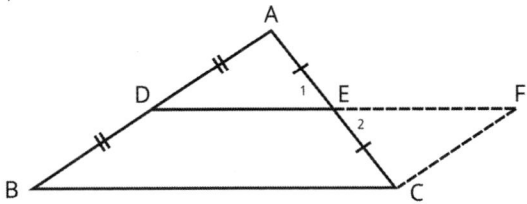

Figure 24.6 Triangle ABC with F projected. Image credit: Ernesto Mora.

Then,
$\angle 1 = \angle 2$ because they're vertical angles.
$AE = CE$, by construction.
$\angle FCE = \angle DAE$ because alternate interior angles are equal.
Triangle AED is congruent to triangle CEF by side-angle-side.
Therefore, $DE = EF$ because corresponding sides of congruent triangles are equal.
$AD = DB$ because midpoints bisect the line.
$FC = AD$ because corresponding sides of congruent triangles are equal.
Therefore, $FC = DB$.
Therefore, BCEF is a parallelogram because FC is both parallel and equal to DB.
Therefore, DE is both parallel to BC and equal to 1/2 BC.

Solution to Exercise 32: Prove the "converse" of the midpoint theorem. That is, if a line segment passes through the midpoint of one side of a triangle and is parallel to another side, then it bisects the third side. (This is not the literal converse of the theorem, which would be: if a line intersecting two sides of a triangle is parallel to the third side and equal to one half of it, then the intersection line is the midpoint of the first two sides.)

In the figure below, P bisects AB, and PQ is parallel to BC.
To Prove: Q bisects AC; i.e., prove AQ = QC.

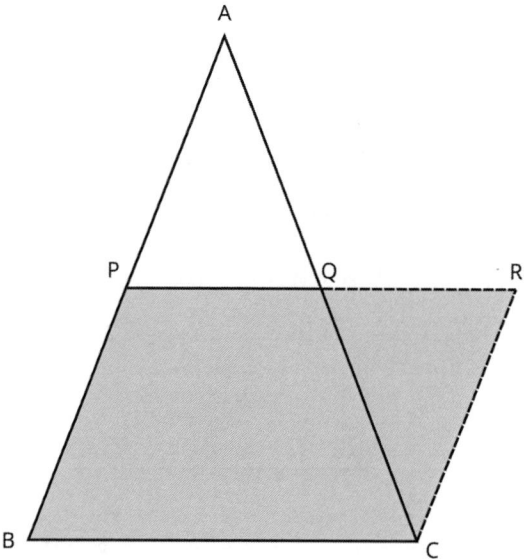

Figure 24.7 Triangle ABC with PQ midpoints and projected R. Image credit: Ernesto Mora.

Draw CR, where CR is parallel to PB and R is the point of intersection of PQ extended.
∠CRQ = ∠APQ because alternate interior angles cut by a transversal are equal.
∠QCR = ∠PAQ because alternate interior angles cut by a transversal are equal.
PBCR is a parallelogram because opposite sides are parallel.
CR = BP because opposite sides of a parallelogram are equal.
CR = AP, since BP = AP.
Therefore, triangle CRQ is congruent to triangle APQ by side-angle-side.
Therefore, AQ = CQ because corresponding sides of congruent triangles are equal.

Solution to Exercise 33: Slightly modify the proof of the theorem called (in the previous exercise) the "converse" of the midpoint theorem to obtain the proof of the literal converse of the theorem.

To obtain a proof of the literal converse of the midpoint theorem, just note that we need an extra step; it's not given that BP = AP. But since PQ is half of BC, it's also half of PR, which means that PQ = QR, which in turn makes triangle APQ congruent to triangle CRQ, which is all we need to finish the proof.

Solution to Exercise 34: Write at least four simple sentences with just P and $\rightarrow$ that use exactly five occurrences of the atomic sentence P. How many of this type are there in total?

The trick is keeping your parentheses straight! Do you have exactly the same number of left parentheses as right ones? If not, you've made some mistake. Here are some example answers:

1. $((P \rightarrow P) \rightarrow (P \rightarrow (P \rightarrow P)))$
2. $(((P \rightarrow P) \rightarrow P) \rightarrow (P \rightarrow P))$
3. $((P \rightarrow (P \rightarrow P)) \rightarrow (P \rightarrow P))$

Here's one last solution, showing how using different types of parentheses can make it easier to read and write:

$$[(P \rightarrow (P \rightarrow P)) \rightarrow (P \rightarrow P)]$$

If you want to stick to one kind of parenthesis, here's a method of numbering your parentheses. The first occurrence of a left parenthesis is followed by a superscript numeral. The first occurrence of a right parenthesis is followed by a superscript numeral corresponding to the left parenthesis being closed. It's easier to see than to read about:

$$(^1(^2P \rightarrow (^3P \rightarrow P)^3)^2 \rightarrow (^4P \rightarrow P)^4)^1$$

The trick is to use these superscripts to keep track of the left and right parentheses that match.

I haven't told you how many of this type of sentence there are in total. Can you guess? If any number of occurrences of P were permitted, there would be an infinite number of sentences.

Solution to Exercise 35: For the connective $\rightarrow$ ("if..., then..."), we'll look for the column where P is T and Q is F (and the rest are T). Select the appropriate column. Then, find the column representing the connective $\vee$ ("or").

P	Q	1	2	3	4	5	6	7	8	9	10	11	12	13	14	15	16
T	T	T	T	T	T	F	T	T	T	F	F	F	T	F	F	F	F
T	F	T	T	T	F	T	T	F	F	T	T	F	F	T	F	F	F
F	T	T	T	F	T	T	F	T	F	T	F	T	F	F	T	F	F
F	F	T	F	T	T	T	F	F	T	F	T	T	F	F	F	T	F

Column 4 represents the connective → ("if…, then…").
Column 2 represents the connective ∨ ("or").

Solution to Exercise 36: Prove that {¬, ∧} can define ∨. Then, prove that {¬, →} is a sufficient set of connectives.

We can define away the connective ∨ ("or") with the following equivalence: define $P \vee Q$ as $\neg(\neg P \wedge \neg Q)$. If it's not immediately obvious, draw yourself a truth table to confirm it. Then, think about it using ordinary English language. "Either P or Q (or both)" is equivalent to "Not neither P nor Q."

I'll leave it to you to prove that {¬, →} is a sufficient set of connectives. But if you're having trouble, here are a couple of hints: $P \rightarrow Q$ is equivalent to $\neg(P \wedge \neg Q)$. It's also equivalent to $\neg P \vee Q$.

Solution to Exercise 37: Prove that {¬, ↔} is not sufficient.

To prove that {¬, ↔} is not sufficient requires a simple observation. Regardless of what's on either side of the ↔, the truth table will always have an even number of Ts and Fs. So, for instance, "or" cannot be represented, nor can "if… then," nor even "and" because their truth tables do not have an even number of Ts and Fs. So, in fact, none of the standard, two-placed connectives can be represented. A nifty observation, isn't it?

Solution to Exercise 38: Show that both the Sheffer stroke and the dagger are each sufficient alone for representing all truth functions.

Sheffer stroke sufficiency:

$\neg P$	$P \mid P$
$P \vee Q$	$(P \mid P) \mid (Q \mid Q)$
$P \wedge Q$	$(P \mid Q) \mid (P \mid Q)$
$P \rightarrow Q$	$(P \mid (Q \mid Q)$

Peirce stroke sufficiency:

$\neg P$	$P \downarrow P$
$P \vee Q$	$(P \downarrow Q) \downarrow (P \downarrow Q)$
$P \wedge Q$	$(P \downarrow P) \downarrow (Q \downarrow Q)$
$P \rightarrow Q$	$((P \downarrow P) \downarrow Q) \downarrow ((P \downarrow P) \downarrow Q)$

Solution to Exercise 39: Prove $(\neg\neg P \rightarrow P)$ with *reductio ad absurdum* (RAA).

Recall that the numbers immediately following the lines of proof are assumption numbers; the actual lines used in the inference rules are on the extreme right, in parentheses.

To Prove: $\neg\neg P \rightarrow P$

1. $\neg\neg P$	1, As (1)
2. $\neg P$	2, As (2)
3. $\neg\neg\neg\neg P \rightarrow \neg\neg P$	1, CP (1)
4. $\neg\neg\neg P$	1, 2 MT (2, 3)
5. $\neg P \rightarrow \neg\neg\neg P$	1, CP (2 is eliminated) (1, 4)
6. P	1, MT (1, 5)
7. $\neg\neg P \rightarrow P$	CP (1 is eliminated) (1, 6)

The strategy here is to arrive at line 3, which is easy because the assumption of $\neg\neg P$ was made first. And the negation $\neg P$ was made on the following line. Then, after $\neg\neg\neg P$ appears on line 4, all is well.

Solution to Exercise 40: A simple computer program could be written showing which fraction (representing a guest in Hilbert's hotel) is paired with which whole number (representing a hotel room). See whether you can write such a program to show this.

You need two variables—well, actually three: d to keep track of the denominator, n for the numerator, and S for their sum. Beginning with $S = 2$, let both n and $d = 1$. Since their sum, 2, is reached, there's nothing more to do (there's only one way $n + d = 2$). Now, let S be 3. There are now only two ways the numerator and denominator can sum to 3: 1/2 and 2/1. Which one goes first? We've decided to make 1/2 the first one and 2/1 the second—i.e., $n = 1$ and $d = 2$. Now that $S = 3$ has been reached in every possible way, increase S to 4. This means that there are now three ways for n and d to sum to 4: 1/3, 2/2, and 3/1. This time around, let d begin as 1 and $n = S - d$ (reversing the previous choice of $n = 1$). The general idea is to keep increasing S by 1 for each pair n, d, while n and d alternately begin with 1. Then the one with the lower value of the two increases by 1, while that with the higher value decreases by 1 until S is reached. Then, increase S by 1 and continue. It's a lot simpler to do than to explain.

Solution to Exercise 41: Prove that an infinite binary decimal series is denumerable, or that it's not.

An infinite series of 1s and 0s is not denumerable. To prove it, use the diagonal argument, like this:

10110001001...
1**1**101000111...
10**0**01111100...
111**1**1100001...
1010**0**110100...

Then, replace all bold 1s with 0s and all bold 0s with 1s. The proof is the same with decimals.

Solution to Exercise 42: Write a Turing machine program to add 2 to 3 in a more complicated fashion. We begin with 0011011100, which represents "2 + 3," and we will end up with 0000011111, which is 5.

The program is to begin at the leftmost 1. Erase it (putting in a 0), then progress right to the next 0. Pass the 0 and keep going right through the three 1s to the next 0. Write a 1 where this 0 was, then go left, pass through the 1s, and also pass the intervening 0 to reach the leftmost 1. Erase this 1 as was done before with the first 1. Then, just as was done with the first 1, go to the right, passing all the 1s as before. And then write a 1, then return leftward until you get to the 0 that is farthest to the right. Then go right for the next 1 to be added. But there is no next 1 because we've added the two 1s that were previously there. So, when the TM program passes the (once) intervening 0 and then goes left for the next 1 to be added, it finds a 0. That means there are no more 1s to be added. So, the TM moves right to the first 1 and halts. It halts on the leftmost 1, which, being a series of five 1s in a row, is what was desired. This program is perfectly general and will add as many 1s as possible to another series of 1s. That is, it's the addition program, which adds any two numbers.

Solution to Exercise 43: Find a True Gödel-type sentence that Smullyan's machine cannot print.

Smullyan's sentence is $\neg PN(\neg PN)$.

References

1. P. Hoffman, *The Man Who Loved Only Numbers*, Hyperion (1998).
2. B. Schechter, *My Brain Is Open*, Simon & Schuster (1998).
3. E. T. Bell, *Men of Mathematics*, Simon & Schuster (1986).
4. Euclid and T. Little Heath, *The Thirteen Books of Euclid's Elements*, vol. 2: Books III–IX, Dover Publications, pp. 412–413 (1956).
5. Aristotle, *Physics*, translation by Robin Waterfield, *Oxford World's Classics*, p. 74 (2008).
6. E. Maor, *The Pythagorean Theorem: A 4,000 Year History*, Princeton University Press (2010).
7. Euclid and T. Little Heath, *The Thirteen Books of Euclid's Elements*, vol. 1: Books I and II, Dover Publications, pp. 338, 354 (1956).
8. T. Little Heath, *A History of Greek Mathematics*, vol. 1: *From Thales to Euclid*, Dover Publications, p. 145 (1981).
9. All puzzles in Chapter 8 are from L. Carroll and W. Warren Bartley III, *Lewis Carroll's Symbolic Logic*, C. N. Potter (1986).
10. L. Carroll, *Complete Works of Lewis Carroll*, Vintage, pp. 819–821 (1976).
11. R. Smullyan, *A Beginner's Guide to Mathematical Logic*, Dover Publications, p. 5 (2014).
12. Game-show problem and responses from https://web.archive.org/web/20181118225309/http://marilynvossavant.com/game-show-problem/
13. N. F. Leopold, *Life Plus 99 Years*, Popular Library, pp. 113–115 (1958).
14. S. Kierkegaard, *Philosophical Fragments; Johannes Climacus*, edited and translation by H. V. Hong and E. H. Hong, Princeton University Press, p. 37 (1985).
15. St. George Stock, *Stoicism*, Archibald Constable, p. 36 (1908).
16. M. Gardner, *Aha! A Two Volume Collection*, The Mathematics Association of America (2006).
17. https://martin-gardner.org/MATHEMATICIAN.html
18. https://martin-gardner.org/Testimonials.html
19. D. Hilbert, "Mathematical problems," *Bulletin of the American Mathematical Society* **8**(10), 458 (1902).

20. Y. Matiyasevich, "My collaboration with Julia Robinson," *The Mathematical Intelligencer*, **14**(4), 41 (1992).
21. https://www.math.uni-bielefeld.de/~sillke/PUZZLES/steiner-lehmus
22. H. Weber, *Leopold Kronecker*, in *Jahresbericht der Deutschen Mathematiker-Vereinigung*, Band 2. Reimer, Deutschen Mathematiker-Vereinigung p. 19 (1893). (There it is reported that Kronecker delivered this remark to a meeting of the Berlin Naturalists' Society in 1886.)
23. D. R. Hofstadter, *Gödel, Escher, Bach: An Eternal Golden Braid*, Basic Books, p. 404 (1979).
24. J. M. Plotkin, "Who Put the 'Back' in Back-and-Forth?" *Logical Methods, Progress in Computer Science and Applied Logic* **12**, 705–712 (1993).
25. P. Geach and M. Black, editors and translators, *Translations from the Philosophical Writings of Gottlob Frege*, Basil Blackwell, p. 234 (1960).
26. J. van Heijenoort, editor, *From Frege to Gödel: A Source Book in Mathematical Logic, 1879–1931*, Harvard University Press, p. 127 (1967).
27. S. Kanfer, *Groucho: The Life and Times of Julius Henry Marx*, Vintage, p. 5 (2001).
28. G. Boolos, "The iterative conception of set," *The Journal of Philosophy* **68**(8), 215–231 (1971).
29. Paraphrased from J. Shoenfield, *Mathematical Logic*, Addison-Wesley (1967).
30. P. Singh, "The so-called Fibonacci numbers in ancient and medeival India," *Historia Mathematica* **12**(3), 229–244 (1985).

 **Charles L. Silver** has a long teaching history. He got his PhD in Philosophy from the University of California Berkeley, studying under Alfred Tarski, Julia Robinson, John Searle, Benson Mates, and Paul Feyerabend. His thesis was an analysis and criticism of Saul Kripke's theory of proper names. He has taught logic, mathematics, philosophy, and computer science at many major institutions, including Berkeley, the University of Memphis, and Smith College. He worked in the private industry for McDonnell Douglas in systems optimization. His previous publications include a logic textbook, *From Symbolic Logic to Mathematical Logic*, and a book on theories of mind, *The Futility of Consciousness*. He has also edited and consulted on several films, among them *Gates of Heaven* (supervising editor), *A Brief History of Time* (about the life and the physics of Stephen Hawking), *N Is a Number: A Portrait of Paul Erdős*, and *Julia Robinson and Hilbert's Tenth Problem*. His interests are in axiomatic set theory, philosophy of logic and language, and computer science. But one long abiding interest is in the history of mathematics, and in introducing problems from the history of mathematics into modern discourse.